First Fruit

First Fruit

The Creation of the Flavr Savr™ Tomato
and the Birth of Genetically Engineered Food

Belinda Martineau

McGraw-Hill

New York   Chicago   San Francisco   Lisbon   London
Madrid   Mexico City   Milan   Montreal   New Delhi
San Juan   Seoul   Singapore   Sydney   Toronto

Library of Congress Cataloging-in-Publication Data

Martineau, Belinda.
  First fruit: the creation of the Flavr savr™ tomato and the birth of
genetically engineered food / Belinda Martineau.
    p.   cm.
  Includes bibliographical references (p.).
  ISBN 0-07-136056-5
  1. Transgenic plants. 2. Crops—Genetic engineering. I. Title.

  SB123.57 .M37 2001
  635.'.642233—dc21   2001018038

McGraw-Hill

A Division of The **McGraw-Hill** Companies

Printed and bound by R. R. Donnelly & Sons Company.
Designed by Michael Mendelsohn at MM Design 2000, Inc.

This book is printed on recycled, acid-free paper containing a
minimum of 50% recycled, de-inked fiber.

To my grandfather

James A. Martineau

for interest in and, hence, fostering of

my scientific and business work

and

to my grandmother

Alice Matravers

for complementarily supporting

my homework

and

in loving memory of

Uncle Ted

Contents

Preface

The U.S. Food and Drug Administration (FDA) held a series of meetings in late 1999 that were designed to gain feedback from the public on the agency's policies regarding genetically engineered foods. As a pioneer in the field of biotech foods, part of Calgene, Inc.'s successful effort to gain FDA approval of the Flavr Savr™ tomato, the world's first commercially available genetically engineered whole food, I had been closely following the debate brewing over them. Public feedback on the topic was sure to be anything but dull. Curiosity, therefore, led me to Oakland, California, on December 13 for the last of the FDA's three meetings, entitled "Biotechnology in the Year 2000 and Beyond." It was an eye-opener.

I was surprised to discover, for example, that although the Flavr Savr tomato had been off the market and unavailable for a couple of years, it still played a prominent part in the debate, especially on the pro-genetic engineering side. Deputy Commissioner for International and Constituent relations Sharon Smith Holston, mentioned it in her opening remarks. James Maryanski, Biotechnology Coordinator for the FDA's Center for Food Safety and Applied Nutrition, in his general description of the FDA's policy toward these new foods, emphasized the lengthy, full review his agency had conducted on Calgene's tomato. The FDA's Food Advisory

Committee, Maryanski explained, not only agreed with the agency's conclusion that the Flavr Savr tomato was safe but also recommended that in light of the lack of safety concerns over Calgene's tomato, subsequent biotech foods need not undergo similarly thorough assessments prior to commercialization.

Maryanski's talk left me feeling anxious and ambivalent. I was proud, on the one hand, that the several years' worth of experiments my colleagues and I had carried out to demonstrate the safety of the Flavr Savr tomato had stood up to the agency's scrutiny so well that they had helped establish the FDA's general process for dealing with all biotech foods. But, on the other hand, I couldn't think of a genetically engineered product more innocuous than the Flavr Savr tomato (which contains minimal foreign material). The fact that Calgene's tomato had paved the way for a process that involved only voluntary consultations between the producers of biotech products and the FDA made me feel, therefore, somewhat uneasy.

The intemperance of the debate added to my unease. As she opened the meeting Deputy Commissioner Holston made what I thought at the time was a rather unusual request of mature adults in a reasonably official setting. She asked everyone to "listen to one another." She knew better than I did. Throughout the FDA's presentations, two expert panel "discussions" on scientific, safety, regulatory, and labeling issues, and nearly 100 2-minute talks by members of the general public, not much meaningful listening seemed to take place between proponents and opponents of biotech foods in Oakland's Federal building that day. The amount of meaningful exchange inside, in fact, appeared roughly equivalent to the amount that occurred between the rival rallies, one fresh from the Seattle WTO protest a couple of weeks earlier, staged outside in the courtyard throughout the day. Nearly nil.

This lack of effective communication on the subject of genetically engineered foods wasn't for lack of desire to exchange information, however, or so it appeared on the surface of the debate. Opponents repeatedly asked for more information, in the form of more safety data and the labeling of biotech food products, while proponents often admitted that a "serious information gap" existed between the two sides. The most consistent message Deputy Commissioner Holston said she had heard throughout the proceedings, in fact, was that "consumers need to be educated."

The problem, I decided as I pondered the situation on my train ride home that evening, was largely in the execution (or lack thereof) of bridging that information gap, especially, it seemed to me, on the part of the pro biotech camp. At the FDA's meeting (and from what I'd seen previously in the popular press) representatives of the ag biotech industry and its supporters in the academic community tended to convey general ideas, as opposed to specific facts, to the public. Proponents rarely referenced safety data published in scientific journals or accessible, through the Freedom of Information Act, in various government documents. And while communicating technical results effectively in public settings is difficult, to say the least, relying on oversimplifications could, I believed, be more troublesome.

And the trouble with the general ideas being conveyed by the ag biotech industry's proponents, as I saw it, was that taken at face value, many of them weren't very convincing. The oft-given description of genetic engineering as an extension of traditional breeding methods, for example, had once been referred to by an acting associate commissioner for the FDA's Legislative Affairs, as the "spin" the

Clinton administration had (back in 1993) put on the matter of biotech foods. "Spin," I felt quite certain, was not the kind of information the opponents of agricultural biotechnology wanted from the new industry's supporters. Neither were they reassured to hear that a so-called consensus supporting the idea that biotech crops were safe to grow on tens or hundreds of millions of acres existed among scientists utilizing genetic engineering under the highly contained conditions of their own labs or small field trials. Opponents were being told that there was "no evidence that any of these products is unsafe" when what they were asking for was positive evidence that biotech foods were, in fact, safe to grow and consume.

That's why I decided to write this book. James Maryanski's speech at the FDA's meeting convinced me (and friends of mine who still work in the ag biotech business confirmed) that Calgene's data supporting the safety of the Flavr Savr tomato remain remarkably relevant. Many of the experiments we did at Calgene, for example, continue to serve as the boilerplates for other companies in assessing the safety of their new products. The Flavr Savr tomato's story includes a (if not the) prime example of experimental data used to demonstrate the safety of biotech foods. I hope, therefore, that telling this story will provide opponents with hard facts they can use to objectively evaluate agricultural biotechnology and, thereby, help move the debate over genetically engineered foods beyond the level of slogans and sound bites. Proponents will be reminded that despite being labeled as such, sales of the world's first genetically engineered whole food were brisk. Labels, therefore, need not necessarily serve as warnings on biotech foods.

In this attempt to move beyond slogans and sound bites, however, I have used the following quote from Albert Einstein as a guiding principle. "Most of the fundamental ideas of science are essentially simple and may, as a rule, be expressed in a language comprehensible to everyone."[1] Human beings also play a critical role in my rendition of the birth of the ag biotech industry. I therefore hope readers of *First Fruit* will find it intellectually nourishing yet easily digestible.

Belinda Martineau

Acknowledgments

Writing and researching this book led to rekindled friendships with a host of former employees, friends, and acquaintances of Calgene, Inc., all of whom enthusiastically helped me reconstruct an exciting, although sometimes nerve-wracking, era in all our lives. Thanks to Aubrey Jones, Bill Hiatt, Dave Schaaf, David Stalker, Don Emlay, Donna Scherer, Jim Metz, John Callahan, Keith Redenbaugh, Ken Moonie, Liz Lassen, Matt Kramer, Ray Sheehy, Rick Sanders, Roger Salquist, Steve Benoit, Toni Voelker, Joe DeVerna, George Bruening, Ron Martin, and Alan Bennett for helpful discussions and/or reviews of various chapters. My apologies for the occasional question that triggered not only a memory but a concomitantly queasy stomach.

I am also indebted to my network of friends and family. Some read early drafts (Pete Martineau, Ed Blakely, Tom Morgan, Jan Goldberg, Michael Plotkin, and Anna Grace), others cared for my children so that I could work (Tracy Chadbourne, Kristin Brandt, and Sheri Smith) and a couple made sure that I was fed during hermit writing sessions (Tim Smith and Marty Tuttle). Thanks to Robert D. Darragh, Teresa Darragh, Mary Martineau, and Angie McIntyre for supplying information and to Anne Darragh, David Ford, Kathryn Krenz, Eleanor Martineau, Mary Darragh, Helen Martineau, and Marcia Lieberman for motivation. Many others, so numerous that it would be too ominous a job to mention them all, politely put up with my sole source of conversation for a solid year. I could not have completed this project without Mona Martineau's reliable grandmothering; thanks, Mom. I am especially grateful to my sister, Pamela Martineau, not only for her considerable editing skills and time spent watching over her local nieces, but also for her consistent encouragement at a time when no one but the two of us thought this project worth pursuing. She was my inspiration before my editor, Amy Murphy, took over the job.

As much as I tried to minimize the effect writing this book had on my immediate family, I was not completely successful. When asked by a schoolmate's parent what her mother did, my older child replied, "She writes chapters." My younger daughter complained more directly, "Will you ever be finished with Chapter Five, Mama?" My husband, for his part, reviewed every line and cared for the kids more than any employed father should ever have to. I will be forever thankful to Antonia, Laurel, and Bob for their support and understanding during this project.

First Fruit

CHAPTER ONE

The Gleam in Management's Eye

In early 1988 the world's first genetically engineered whole food, the Flavr Savr™ tomato (or, rather, its prototype), was—literally and figuratively—barely alive. Yes, young tomato plants containing the Flavr Savr gene were alive and well in the Davis, California, greenhouses of Calgene, Inc., the independent agricultural biotechnology company where the high-tech tomato was being developed, but Calgene's management was not ready to stake the company's reputation or future on the small red orbs those plants were producing. In fact, Calgene was moving away from tomato as a business opportunity altogether. In an article published in the local Davis paper in March 1988,[1] mention of the project that would later be dubbed Flavr Savr was barely made by Roger Salquist, Calgene's colorful CEO. Instead of hyping what would become his company's most famous—and infamous—product, the ex-submarine officer and former Stanford yell leader cheered for Calgene's herbicide-resistance program and the Bromo-Tol™ gene. The BromoTol gene could provide crop plants

1

like cotton resistance to an herbicide that would normally kill them as well as a broad spectrum of weeds. Since Calgene had recently acquired the nation's second-largest cotton seed company, Stoneville Pedigreed Seed, all indications were that Calgene science and Calgene business were coming together nicely to produce and market what was expected to be the company's first genetically engineered product: BromoTol cotton seed.

The implications of the business emphasis on cotton for tomato research at Calgene were impressed on me in a meeting that spring with Robert Goodman, Calgene's Vice President of Research and Development. Goodman opened the meeting by warmly welcoming me into the Calgene family. I had been informed—warned might be a better way to describe it—about this family atmosphere already. During my 14-hour interview some 6 months earlier a high-level scientist, Luca Comai, had asked if I had any trouble standing up for myself because, while Calgeners would stand united in public, they were quite capable of tearing one another's work to shreds in-house. Of course, all was usually forgiven over beers at the local brew-pub on Friday afternoon. I soon learned that this family feeling stemmed partly from the fact that many, even most, of the company's employees on the business as well as the science side had yet to begin their own families. Calgene was a small, close-knit group of people growing up and building the company together. (When individual Calgeners did start having children, the floodgates really opened up. Bob Goodman told me the caterer for the company's 1989 picnic asked him, "What, besides kids, does Calgene produce?")

But, despite the fact that he had just increased the number of Ph.D. scientists on his tomato team by 50 percent — that is, he had hired me, bringing the total number to three—Goodman predicted that "in 6 months time Calgene would be out of tomato research completely." I was shocked. Were all ag biotech businesses run this way, I wondered, beefing up the staff on a program one month and considering the elimination of the same program a few months later? And, if so, was it too late for me to bail out and land an academic job?

But things were about to change again at Calgene, and change in a big way. The very opposite of Goodman's prediction came to pass. By August, not only was tomato research back in vogue, but the company had also announced its intention to enter the fresh-market tomato business. In fact, on August 16, 1988, Calgene Chief Financial Officer Daniel O. Wagster II described the Flavr Savr tomato project as "neck and neck with herbicide-tolerant cotton . . . in Calgene's efforts to get genetically engineered products to market."[2]

What prompted this dramatic turnaround, a turnaround that marked the turning point in the history of the company, indeed, of the whole agricultural biotechnology industry? What changed during those few intervening months? An exciting scientific discovery was interpreted in the context of disappointing financial results. It proved to be a potent combination, one that led both to Calgene's greatest success—the marketing of the world's first genetically engineered whole food—and, paradoxically, to the company's eventual demise.

The Discovery

The discovery came as the result of an experiment that was incredibly simple, especially for a high-tech outfit like Calgene. A few ripe tomatoes were harvested from plants into which the Flavr Savr gene had been inserted through genetic engineering. Ripe tomatoes were also harvested from nonengineered tomato plants for comparison. Both groups of tomatoes were left at room temperature and observed over time. After 3 to 4 weeks the nongenetically engineered tomatoes were noticeably shriveled and rotting, while the Flavr Savr tomatoes looked and felt essentially no different than just picked. As Wagster told the *Sacramento Union*, "tomatoes that have been tested with the synthetic gene last up to four weeks in a non-refrigerated state."[3] In fact, he explained to the *Sacramento Bee*, "It's unclear as to how long it will take until [the tomatoes] do soften."[4] What was clear, especially since the results were extremely photogenic, was that the shelf life of the Flavr Savr tomato was dramatically longer than that of ordinary tomatoes. A picture—worth a thousand words—of Flavr Savr tomatoes alongside their non–genetically engineered rotting cousins was included in Calgene's 1988 and 1989 annual reports. Indicative of the exciting, even sexy, status of the shelf-life experiment, the photograph published in the 1989 report was made a centerfold.

The scientist responsible for this exciting result, the one in charge of all tomato research at Calgene at the time, was William R. Hiatt, an enthusiastic, quick-thinking microbiologist. Bill Hiatt joined the company in 1981 as its first Ph.D. scientist and had been working on the Flavr Savr tomato since the originator of the project, David Stalker, had convinced him to take it off his hands in 1984. At first, Bill felt the project had too slim a chance for success. He only took it on because David, one of Calgene's top three scientists (along with Bill Hiatt and the head of canola research, Vic Knauf), successfully wooed him with the assurance that good science—that is, lots of papers— would come of it. So

Bill hired a molecular biologist as his research associate, and together they constructed the so-called Flavr Savr gene (named, to the chagrin of the tomato science staff, years later by Roger Salquist).

The Flavr Savr gene consisted of a copy of a well-studied tomato fruit gene that coded for an enzyme called polygalacturonase (PG for short), involved in the fruit-ripening process. The primary difference between the PG gene present in unadulterated tomatoes and the engineered gene was that, in the Flavr Savr version, the genetic information that led to production of PG protein had been flipped upside down and backward. The resulting "antisense" PG gene, by some unknown mechanism, shut down native PG protein production in the engineered plants. Apparently, eliminating PG protein in tomato fruit by expressing the Flavr Savr gene increased the shelf life of ripe tomatoes, scientifically an exciting result.

But, like many scientists, Bill was reluctant to publish, in the newspapers or anywhere else, experimental results that he deemed preliminary. In fact, Bill was more reluctant than most. It would have been one thing to talk about how the Flavr Savr antisense PG gene had shut down PG production in genetically engineered tomatoes; that result was no longer preliminary. Bill's manuscript describing that work was accepted for publication in the *Proceedings of the National Academy of Sciences*, a prestigious scientific journal, coincident with Calgene's press release in August 1988. But the shelf-life result was another story. For one thing, the experiment had been carried out only once. It was nowhere near ready to publish. (It would be nearly another 4 years before the shelf-life experiment was published in a scientific journal.) So when reporters and television crews descended upon Calgene in their quest to get the word on the new, long-lived tomatoes straight from the man who invented them, Bill looked for other scientists, any other scientists (whether working with him on the Flavr Savr or not) who would consent to do an interview in his place. He cornered me that August, for example, and, although I was not officially working on the Flavr Savr tomato project yet, enthusiastically encouraged me to speak with a television reporter. "Look at it as an opportunity," he said, confining me in the gaze of his piercing blue eyes. I declined. His hopeful grin gone, his disappointment obvious, he then used the occasion to give me a general piece of advice, "In my opinion, talking to the press is committing professional suicide."

This mercurial style was typical of Bill. He could change his mind, in light of new information, faster than anyone I've ever known. The scary thing was that, according to Bill, his brain used to work even faster. "When I was a kid," he told my office mate and me one day when he was moving in particularly high gear, "I couldn't even walk down the street or go into a restaurant; I couldn't handle all the stimuli coming at me. So," he said and paused as if for emphasis, "I decided then and there to kill off a quarter of my brain cells." As he darted out the door we were left to contemplate whether he had carried out his plan.

In light of his opinion about talking to the press, it was not surprising that Bill Hiatt was not extensively quoted in the newspapers. On the other hand, it was also not surprising that Daniel O. Wagster II was. Dan Wagster was the quintessential preppie; he seemed to revel in the label. "Did you go to a prep school?" I asked, just trying to make small talk one of the first times I met him. "You've got to be kidding," he laughed. "Just look at me." His prep school education was evident in the way he dressed, the way he acted, and especially the way he talked. He was very verbally oriented. This, plus the fact that he, along with Andrew Baum, Vice President of Operations, headed Calgene's business planning effort, made Dan a natural choice as primary spokesperson for the Flavr Savr tomato. Roger Salquist tapped him for the job. And, as pitchman for the project, Wagster did not hesitate to outline the business opportunity Salquist envisioned for Calgene's "slow-to-rot" tomatoes. "There are places in the [$4 billion U.S. fresh tomato] market where premium tomatoes sell for two-and-a-half times the normal price at the wholesale level," Dan Wagster said. "That's clearly where we'd be positioning the product."[5]

The Positioning

"Premium tomatoes" are high-quality, vine-ripened tomatoes. Flavr Savr tomatoes would make good premiums because they could be vine-ripened more easily than traditional tomatoes, "officials at Calgene" explained to the *Sacramento Union*. The Flavr Savr gene would "help block the rapid deterioration of vine-ripened tomatoes. That would permit tomato growers to allow the fruit to remain on the vine until maturity instead of picking them at the immature, green stage, as is the current practice for tomatoes sold in retail stores."[6] And it went without saying (almost) that mature, vine-ripened tomatoes were vastly better tasting and therefore deserving of the "premium" label, than other tomatoes available in grocery stores. Although Dan Wagster told the *Union* that further testing still needed to be done, it seemed the Flavr Savr tomato's destiny had already been determined as a "premium, tastier" tomato.[7]

A tomato that stayed firm enough to naturally ripen on the vine and still survive shipment to market was an extremely exciting prospect, almost too good to be true. In fact, the headline for one newspaper story read: "A store-bought tomato that actually has taste?" But it made a lot of sense, not only for a fresh-market tomato business, but, even more important, to American consumers.

It made sense to the tomato industry because surviving shipment to market was the name of the game for fresh-market tomatoes. Plant breeders had worked for decades to toughen tomatoes and improve their shippability. Shippability was also the reason why tomato producers picked the fruit before it had a chance to start the ripening process, when they were still extremely firm but also completely green. So green, in fact, that Roger Salquist couldn't believe it when he first viewed a videotape of the harvesting process. "It's the wrong god-damned tape," he cried, "those are apples!"

Hard, green tomatoes survived transportation to distributors and grocery stores, where they were artificially ripened using ethylene gas. Fruit harvested with even a speck of color—red, pink, or orange—did not survive the trip. They tended to ripen and soften during transport to a point of uselessness referred to in the business as shrink. Therefore, firmness and its unfortunate associate, greenness, were considered absolutely necessary qualities for fresh-market tomatoes. Without them, high-volume sales over a large geographical region were impossible.

But "gassed green" tomato firmness came with a high price. Those tomatoes looked red and ripe (sort of) on the produce shelf but tasted awfully green once they got home. And the reason that they tasted like cardboard was that they weren't allowed to undergo the normal ripening process on the vine, during which they attained their full flavor. American consumers were unhappy with these artificially ripened tomatoes. The taste of gassed green fruit just didn't come close to the taste of a vine-ripened homegrown tomato.

The Opportunity

Therein lay the real beauty of the Flavr Savr tomato. If tomatoes could be made firm enough to withstand shipment and yet be allowed to ripen on the vine . . . wow, you're out of the gassed green tomato business and into a potentially huge vine-ripe business. For, despite their complaints, each American consumer bought an average of 17 pounds of fresh tomatoes every year. Imagine how many they might buy or what share of the $4 billion total fresh tomato market a business might occupy if a truly good-tasting tomato was readily available. In fact, consumers might not only buy more, but also pay more, for the vastly superior taste of a vine-ripened tomato.

The Flavr Savr tomato also seemed a good bet, in part because of its consumer appeal, for getting around what looked like its only negative feature: the fact that it was developed using genetic engineering. To a public leery of this new technology and the cloning it involved, it might take a pretty compelling reason to get them to buy and eat genetically engineered food, especially the first genetically engineered whole food ever. A flavorful, vine-ripened tomato available year-round (as Calgene would later promise) might be compelling enough. And the fact that the genetic modification itself—the addition of a relatively small piece of DNA containing a known, albeit "antisensed," gene from the same organism—was minimal was also viewed as a plus. A plus not only in easing acceptance by the public, but also for favoring the "likelihood of success" with the U.S. regulatory agencies[8] from which Calgene would seek approval for commercialization of the Flavr Savr tomato. In fact, consumer disappointment with traditionally produced "gassed green" fresh tomatoes was so widespread that, it was hoped, a successfully introduced Flavr Savr tomato could open the door for the entire agricultural biotechnology industry.

Not only was it a product with great consumer identity and appeal, the Flavr Savr tomato also made especially good, and timely, business sense for Calgene. Like most biotech start-ups, the company had steadily bled money early on while its scientists developed their basic technologies. From its incorporation in 1980 through 1986, when Roger Salquist took it public (2,535,000 shares of common stock with net proceeds of $32.5 million) based precariously on his vision of the company's potential, Calgene had been in the red. Before the IPO (initial public offering), the company's reputation had been based almost exclusively on its scientific leadership in plant biotechnology, that is, on its research and development organization. After the IPO and especially during 1987, Calgene's first full year as a publicly traded company, the emphasis turned to business strategy. By 1988, a general business concept that focused on both "input" businesses (e.g., sales of genetically engineered seeds at premium prices) and "output" businesses (value-added processed agricultural products) in core crops (those in which Calgene had a technology leadership position) had solidified into a "vertical integration" strategy.[9] Vertical integration implied control, from farmer to customer, of every business Calgene entered. It was designed to maximize financial returns on future genetically engineered products.

Implementation of the vertical integration strategy commenced immediately. By the end of fiscal year 1988, Calgene had established or acquired businesses on the input (AmeriCan Pedigreed Seed Company) and output (Agro Ingredients, Inc.) sides of Salquist's favorite core crop, the crop around which his vision for the company evolved: canola. To complement its Stoneville Pedigreed Seed Company, Calgene acquired Feffer Delinting and Seed Treating Company in 1987, which would provide a marketing outlet for its cotton genetic engineering projects. But tomato had been an especially tough business nut to crack. Essentially all of Calgene's tomato research, to the tune of a million dollars a year, was carried out under contract with the Campbell Soup Company. The terms of that contract specifically excluded Calgene from going into the tomato business, any tomato business. Hence the downplaying of potential tomato products by Salquist and research by Goodman in early 1988.

If there were any chance that Calgene could buy its way out from under the restrictive contract it had with Campbell Soup, it had to be in fresh-market tomatoes. Campbell was interested in marketing fresh fruits and vegetables, in fact, was experimenting with its Fresh Chef line at the time, but processed foods were still far and away the mainstay of the company. If Calgene restricted its business to fresh tomatoes only, as opposed to tomatoes processed into paste or soup, Campbell might not consider it direct competition and go for the deal. Positioning the Flavr Savr as a fresh-market tomato, therefore, brought vertical integration for a tomato business into the realm of the possible for Calgene. Accordingly, by the time the first descriptions of the Flavr Savr tomato had hit the newspapers, the two companies were already "negotiating a commercialization agreement."[10]

Assuming that agreement could be reached (and it was), the Flavr Savr tomato seemed to have everything going for it. Technically, genetic engineering of tomato was more straightforward than that of either cotton or canola. Genes could be isolated from and reinserted into tomato more easily than with either of these other core crops. Expectations were that, because the development time for a genetically engineered tomato product was faster, the Flavr Savr tomato could beat BromoTol cotton to market.

The Flavr Savr tomato had it over BromoTol cotton in another way, too—politically. While Jeremy Rifkin, President of the Foundation on Economic Trends, and various environmental groups had largely ignored agricultural biotechnology up until that point, the one issue they were up in arms about was herbicide tolerance. Making crop plants resistant or tolerant to herbicides would increase the overall use of these chemicals, they insisted. Farmers would overuse them if they knew that their crops were impervious to them, and the excess would run off the fields and into human and animal water supplies. Al Gore mentioned Calgene's BromoTol cotton specifically in his book *Earth in the Balance.* He described bromoxynil, the herbicide to which BromoTol cotton plants were tolerant, as a "reproductive toxin thought to pose hazards to farm workers."[11] Although Calgene refuted these claims, bad press straight from the White House was hard to overcome.

And one of the biggest pluses of the Flavr Savr tomato was the size of the U.S. market for fresh tomatoes. At $4 billion, the fresh tomato market was at least ten times larger than the potential market for any of Calgene's other genetically engineered projects. A reasonable slice of that pie would be large enough to satisfy Wall Street and Calgene's investors, who expected a big return on the money they'd spent to bail out the red ink–laden company.

The Coincident Financial Disaster

It was especially fortuitous for Calgene to have a big money opportunity like the Flavr Savr tomato in August of 1988, since financial disaster had befallen the company nearly coincidentally with the discovery of extended shelf life in Flavr Savr tomatoes. After 7 years with nothing, at least in terms of a genetically engineered product, to show for itself, Calgene had finally, by virtue of sales of conventional products by Stoneville and Agro Ingredients, made a small profit in 1987.[12] In response to this relative prosperity, the science and support staffs had been beefed up in fiscal 1988 (it was during this spree that I was hired), and land was purchased and elaborate plans were made to build new greenhouses. Then, very suddenly that summer, as the third-quarter books were being closed, prosperity was gone and the company plunged into a depression. Two million dollars in losses were recorded during that quarter alone. The greenhouse plans were quickly scrapped, and layoffs were rumored.

Calgene's dreadful financial status was mentioned, but only peripherally, in the newspaper articles announcing the Flavr Savr tomato that August. The startling losses were primarily attributed to "large expenses in research and development."[13] An impending layoff of some 10 percent of Calgene's employees, primarily among the science staff, wasn't mentioned at all in any of the articles. The "super tomato" had given Calgene management something exciting and positive to report on instead of the gloom and doom that was actually occurring at Calgene at the time. So, in addition to all its other attributes, the Flavr Savr tomato also had great timing. In fact, there was only one problem with what appeared to be the perfect flagship product for Calgene, the Flavr Savr tomato, in August 1988. It didn't exist. It was only a dream.

The Dream

"Officials at Calgene" had taken a bona fide scientific breakthrough, the extended shelf-life phenomenon, and, through a kind of logical extrapolation, invented the Flavr

Savr tomato. They assumed that, because Flavr Savr tomatoes stayed intact longer than regular tomatoes after they'd been harvested ripe (the results of the simple shelf-life experiment), they should be firmer as they ripened on the vine as well. If the genetically engineered tomatoes were firmer than their nonengineered counterparts during the ripening process, then they could be picked "vine ripe" and still survive shipment to market. And, because they ripened on their own, those tomatoes should also taste better than standard fresh-market tomatoes that were picked while still green and artificially ripened with ethylene gas.

The reasoning made sense. It was logical. And there is nothing wrong, per se, with a dream born of deductive reasoning. But in scientific parlance the Flavr Savr tomato dream was little more than a hypothesis, and a hypothesis was only the beginning, the start of an experiment, "a tentative assumption made in order to draw out and test its logical or empirical consequences."[14] In August 1988, Calgene was still a "story stock," its value based not on its products or sales but on its story. The Flavr Savr tomato, a "tentative assumption," was Calgene's story.

Calgene's executive staff obviously had a very positive outlook on that tentative assumption. Perception leads to reality in business, and all that stood between their perception and the reality of a "premium, tastier" variety of tomatoes was simply "further testing."[15] And although Calgene's legal right to enter a tomato business had yet to be obtained from the Campbell Soup Company, Salquist and his business planning staff were already "considering purchasing other businesses involved in growing and handling fresh tomatoes."[16] Calgene business was bullish on the Flavr Savr tomato.

Those of us in Calgene's scientific trenches were not as optimistic about the Flavr Savr tomato as the business planning staff was. It's not that we were a pessimistic lot (although some of us were). It's just that we were considerably more realistic. If the description of the Flavr Savr tomato had been limited to its ability to linger on the grocer's and/or the consumer's shelf weeks longer, it would have been one thing. Albeit preliminary, at least there were some experimental data to support the existence of that tomato. But a vine-ripened fresh tomato that could survive the shipping process? That tomato was purely hypothetical. It was an untested prediction, an extrapolation of the shelf-life results and, as such, was something that we, as scientists, would only mention in our notebooks or during discussions at a project meeting. We might make bets with other scientists as to the outcomes of experiments planned to test our predictions, but splash them in the newspapers? Never. We knew from our scientific training that testing the prediction, the painstaking process of drawing out its "empirical consequences," lay between us and an actual fresh tomato product. What's more, we knew from years of experience that empirically derived consequences don't always support one's original hypothesis. There was a chance that the Flavr Savr tomato hypothesis would not pan out. That was the scientific reality of the situation.

The Reality

But scientific reality was not the relevant issue. The company was in sell mode; before it could develop and sell a genetically engineered tomato, it had to sell itself as a viable business. And with the advent of the Flavr Savr tomato, Calgene's business plans began to affect its science staff. Prior to the conception of the Flavr Savr tomato dream and within the limits of general project targets, Calgene scientists had been left largely to their own devices and had done excellent scientific work, as evidenced by their publications in some of the country's most prestigious scientific journals. The company's reputation, not to mention its patent portfolio, rested largely on that science. The most anti-industry professor I knew at U.C. Berkeley, Michael Freeling, gave Calgene a thumbs-up based on its scientific reputation in 1987. Professor Freeling's opinion of the company was one of the main reasons I accepted a job at Calgene. But in August 1988, scientific independence at Calgene began to wane. Interaction with, as well as direction from, the business staff became inevitable.

The impact of Calgene business on its science became painfully clear when the layoffs of 1988, the second in the company's history, finally occurred. Wagster displayed his characteristic confidence as he explained in-house why R&D spending was to blame for the fiscal 1988 losses. "Wall Street understands that research and development is expensive," he told the crowd. And layoffs were necessary to demonstrate to Wall Street that Calgene's leaders "could make the tough decisions."

It was a hard line for those of us in the audience at the time to swallow. From our vantage point, and apparently Salquist's based on his report to shareholders in the annual report that year,[17] the company's management team had not achieved its financial targets and yet the science staff was paying for it. And Wagster, as the company's CFO (and formerly the Manager of Accounting Controls at IBM), made an especially disagreeable bearer of the bad news. The conspicuous display of his brand-new, top-of-the-line Jeep Grand Cherokee at work the afternoon of the "downsizing" only worsened his relationship with what remained of the R&D staff.

The new business era was not viewed negatively by all Calgene scientists. After all, for the company to survive, it had to make money. Calgene had to take—and make—that leap from venture to bona fide business. But there were those who felt the vertical integration method of building the business might not be the way to go. David Stalker was the most vocal among them. Rather than try to take on several different businesses from the ground up, as would be required with the vertical integration strategy, he thought that Calgene should stick with what Calgene did best: cloning genes and transferring them to and expressing them in plants. Those cloned genes and/or transformed plants containing them could then be sold or licensed to established agricultural businesses and thereby incorporated into already going concerns.

It sounded like a reasonable course of action. The prevailing sentiment among Calgene's business elite, however, was that there was not nearly as much money to be made in such a "gene boutique" business as with the "vertical integration" strategy. And although Stalker was highly respected for his science, as far as I was aware, no one on Calgene's executive staff considered the gene boutique business idea seriously.

Bill Hiatt sided firmly with the vertical integrators. He was convinced, perhaps under the influence of what John Kenneth Galbraith refers to as the "peculiar magic that is thought to be at the command of those intimately involved with financial matters,"[18] that the company had to make the kind of money that only vertical integration could supply. A gene boutique business simply could not fit the required "big money" bill. So, partly due to his philosophical trust in Calgene's financial gurus and almost certainly partly due to his reservations about dealing with the media, Bill Hiatt stood in the background as "Calgene officials" whipped up a frenzy around his Flavr Savr tomato project.

As a result, the Flavr Savr tomato newspaper articles all reflected an unwavering optimism. And in the reflection of that optimism, the Flavr Savr tomato's future looked very bright, not to mention straightforward. "Our objective now is to make [tomatoes] sturdy and good-tasting" Wagster told the *Sacramento Union.* [19] Of course, the "our" in Wagster's quote meant Bill Hiatt. In more ways than one, Bill had his work cut out for him.

CHAPTER TWO

Scientific Conception

(amid Misconceptions and Controversy)

During the media blitz in August 1988 that announced the Flavr Savr tomato concept to a tasty tomato–starved world, Bill Hiatt had other matters on his mind. Public relations were all well and good for vice presidents. But Bill Hiatt was first and foremost a scientist, and a scientist, even an industrial scientist, feels pressure to publish or perish. Publishing his work in peer-reviewed scientific journals—not the newspapers—was still his primary goal.

Publish (and Patent) or Perish

Bill's most important long-term publication goal was to provide hard scientific evidence that the Flavr Savr tomato concept was realizable. To hypothesize that shutting down a gene correlated with tomato softening would make vine-ripened tomatoes firm enough to survive transportation to grocery stores in an economically efficient way was one thing. Demonstrating that this scenario was in fact feasible was another matter entirely. The fact that this critical demonstration depended on an infamous 20-year-old "squashing" technique for measuring tomato firmness didn't help Bill sleep any better at night. He set his best people about the daunting task.

In the short term, Bill had a chance at being first to report in the scientific literature that it was possible to shut down specific genes in tomatoes using genetic engineering. This result was what provoked the Calgene media blitz in the first place. Being first to publish in a scientific journal was significant because, at least theoretically, every scientist whose work further elucidated the landmark first result (and if it was truly landmark work there would be many) would refer to Bill's paper when publishing his or her own results. Publishing first was a means to becoming famous, at least among your scientific peers.

Luckily for Bill, Calgene had been fashioned after the biomedical biotech companies in terms of encouraging rapid publication.[1] Publication was good for biotech business because it established "prior art" and thereby prevented competitors from obtaining patent protection of the published material. Prior art made an idea "obvious" to anyone "skilled in the art" and, therefore, nonproprietary. It was an effective defensive strategy. Publication also allowed industrial scientists to participate in several of the traditions on which academic research is based: building on previous efforts, duplication of others' work, and, most important, peer review.[2] Consequently, it not only kept the scientists already on board at a company happy but also served as enticement for academic scientists to join the industrial ranks. Publication in high-quality, well-known journals, moreover, served as good advertising for young biotech companies looking at years of research and development before potential products would make it through the pipeline. Scientific papers were product placeholders in the early days of ag biotechnology.

Of course, the offense that countered the publication defense was prior patenting. As was true at most other biotech companies, patent filing was a prerequisite for publishing at Calgene. The last thing the company wanted was for any of its scientists to publish or even publicly speak out about their potentially patentable results before they had filed for patent protection. Any public release of an individual's own material also constituted "prior art" for an idea. To use a basketball metaphor, publishing before patenting could block your own team's shot at the basket.

Bill, along with several of his Calgene colleagues, had fulfilled his patent obligations by August 1988 by filing two patent applications. One requested patent coverage of the tomato gene sequence that was crucial to the Flavr Savr tomato project, the other coverage of the use of anti-sense technology to regulate any gene's expression in plants. He had sent the U.S. Patent and Trademark Office final experimental information demonstrating, he hoped, that his inventions had been "reduced to practice" that very month. Confident that it was now impossible to block his own shot at patent protection, Bill had also submitted his results for publication. His receipt of word from the U.S. National Academy of Sciences that, yes, the Academy would publish his manuscript in its *Proceedings* journal had coincided with Calgene's first Flavr Savr tomato press release. So, in August 1988, with the patent situation apparently under control and his (supposed) landmark paper accepted for publication, Bill appeared to be sitting pretty. I believe, however, he might actually have felt more like a sitting duck.

The Competition

Bill knew he had stiff competition on the Flavr Savr tomato project, and he knew, based on rumors that had surfaced at a Gordon Research Conference on plant senescence the previous month, it was right on his tail. Filing a U.S. patent application was one thing, but having that patent actually issued was quite another. Because U.S. patent applications are not published until they are issued, it was impossible to know whether or when another group had filed a similar and/or possibly blocking application. Bill knew that a group led by Don Grierson at the University of Nottingham in Great Britain was also trying to improve tomato fruit quality with its own version of the Flavr Savr gene. Had Grierson applied for a patent on the process? Unfortunately for Bill, who was not good at waiting, only time would tell.

Of perhaps even more concern to Bill that August was the fact that his scientific reputation was on the line. Sure, his Flavr Savr tomato manuscript had been accepted for publication, but when would it actually appear in print? Based on the Gordon conference rumors, Don Grierson and his group had submitted a paper to the prestigious English journal *Nature*. Would their paper be published before his? Bill had already been soundly beaten to publication in a three-way race to clone the tomato DNA that was to become the Flavr Savr gene. He had no desire to be beaten again.

The Race to Clone
the PG Gene

The race Bill had already lost began in 1984. Three scientific groups, including Calgene's, wanted a copy of the gene responsible for production of tomato fruit polygalacturonase (PG). PG protein was found in large quantities in ripe tomatoes but not in hard green ones. It appeared just as the fruit started to soften and accumulated to higher and higher levels as it got softer and softer. This correlation was one reason why PG had long been implicated as a major contributor to tomato softening.[3] PG was also implicated in the softening process because tomatoes that didn't accumulate PG protein properly, mutant tomatoes, for example, didn't soften properly either. And tomato processors knew that if they didn't carry out a "break" step, that is, processing fruit at temperatures of 73 to 95°C explicitly to deactivate so-called pectolytic enzymes like PG in harvested tomatoes, tomatoes would "soften" into a watery mess not at all conducive to ketchup or other processed tomato product preparation. For all these reasons, scientists like Bill Hiatt and tomato-processing companies like Campbell Soup wanted to get their hands on the gene responsible for PG protein production so they could try to control tomato softening. The Flavr Savr gene project was Calgene's design for a solution to the tomato PG protein problem.

The goal of Calgene's project sounded simple enough: shut off the gene coding for PG protein in tomatoes. However, to use genetic engineering to accomplish this "simple" task required three major steps. Obtaining a clone of the PG gene from an existing tomato plant was just the first step. Second, the clone had to be reinserted into the DNA of isolated tomato cells and those cells regenerated into a complete, healthy tomato plant. And third, the isolated gene copy had to be manipulated in such a way that, upon its reinsertion into a recipient plant, it interrupted the production of PG protein not only by itself, but also by the native, nonengineered PG gene resident in the recipient plant.

But in 1984, no one had cloned a PG gene. No one had successfully inserted, via genetic engineering, a gene of any kind into a full-grown, fertile tomato plant. And no one had ever used antisense technology, the method chosen by Bill and his colleagues, to curtail the expression of any plant gene. The cards seemed highly stacked against the success of the Flavr Savr gene project.

Undaunted by the odds and spurred on by the signing of a contract with Campbell Soup, Bill set about to accomplish task number one, cloning the gene that was responsible for producing PG protein in the tomato. Simultaneously at Calgene, a group led by JoAnne Fillatti addressed task number two, developing a method of transforming (inserting genes into) and regenerating tomato plants. And, while manipulating a cloned PG gene, once obtained, into an antisense configuration would be a straightforward genetic engineering procedure, whether that reconfigured "antisensed" gene would, in fact, shut down PG protein production in tomatoes was anybody's guess. That was simply the reality of genetic engineering projects at Calgene at that time and for at least the next dozen years. (More conventional plant scientists who conduct research with herbicides refer to their "anybody's guess" experiments as "spray and pray." The corresponding category for genetic engineers could be dubbed "clone and moan.") But Bill and the management teams at both Calgene and Campbell Soup were cautiously optimistic. Bill hired a research associate to help him clone the PG gene.

Raymond E. Sheehy received his B.S. degree in bacteriology from the University of Idaho, Moscow, and his M.S. in molecular genetics from Washington State University, Pullman. Ray had considerable cloning experience and was working at Oregon State, Bill Hiatt's alma mater, in 1984. Bill hired Ray to clone the tomato PG gene.

Although the two men were quite different temperamentally, they made a good working team. Ray moved slowly and deliberately. He was shy, at least professionally. He usually spoke up in meetings only when directly addressed and then only reluctantly. Bill, on the other hand, always had some part of his body or another in constant motion, and he always seemed to be in a hurry.

Bill used his sense of humor to get his point across, and Ray, like many other researchers at Calgene, didn't always know how to respond. It irked Bill, for example, that Ray's scientific results, his films, and filters, as well as his notebooks, were "messy." Although Ray had a knack for coming up with a scientific "answer" in a hurry, it was usually in a condition that was far from "publishable." Bill teased Ray about this on a regular basis. Ray responded by either ignoring his boss or emphatically whining, "Bill!" Ray continued to take Bill's admonitions lightly, even when they came up during his yearly performance reviews, until the day his notebooks had to be scrutinized as part of a patent dispute. Bill and Calgene's in-house attorney, Liz Lassen, met with the entire research staff and reminded us to avoid the use of question marks and phrases like "Where am I?" and "What was I working on?" in our notebooks. No names were mentioned, but it seemed reasonable to assume that, since he had done the lion's share of the work under dispute, a few of these phrases might be found in Ray's notebooks.

Bill also teased Ray about a defective gene he possessed. Ray was red-green color blind and consequently had a very difficult time differentiating tomatoes in various stages of ripening, a task essential to his job. "From now on," Bill told him, "before we hire anyone for this program, they'll be screened for color blindness."

Ray and Bill, like most Calgene employees in the early days, put in many long hours, working evenings and weekends, chasing down and cloning the PG gene. Calgene often had as many cars in the company parking lot on a Saturday as on a weekday. Bill and Ray worked so hard for several reasons. For one thing, it was important to impress Campbell Soup in order to keep those contract revenues rolling in. Ag biotech was barely off the ground, and Calgene wanted to keep its few clients happy. The sheer magnitude of the job at hand also served to light a fire under the two scientists. The primary reason, though, was that time was of the essence. Bill and Ray could feel the competition biting at their heels.

Don Grierson's group at the University of Nottingham was cloning as many tomato fruit-ripening genes as they could get their hands on.[4] At the time, they hadn't identified any of them, but, because PG protein was found in ripening tomatoes, the gene coding for it was expected to be among Grierson's "unknown gene" clones. Bill knew it was only a matter of time before Don and his colleagues figured out which of their clones coded for PG.

Alan Bennett led the other group competing with Bill and Ray to clone the PG gene. He was carrying out his studies at U.C. Davis, right in Calgene's backyard. The race was on.

The U.C. Davis PG Cloners

Not only close geographically, Alan Bennett's and Bill Hiatt's methodologies for cloning the PG gene were amazingly similar. Both (in fact all three) groups utilized the PG protein itself to isolate the gene responsible for producing it. This approach was possible partly because there was just so much PG protein found in ripe tomato fruit.*

It was therefore relatively easy to isolate from tomatoes, and it could be separated from other tomato proteins based on its size and enzymatic activity. (PG protein, a pectolytic enzyme, breaks down pectin, a component of the "meat," or walls, of a tomato. Adding pectin to a test tube containing PG protein and then monitoring the loss of pectin served as a test for the presence of active PG protein.) Once isolated, PG protein was injected into a rabbit or other animal with a suitable immune system in order to elicit the animal's immune response. "Booster shots" of the foreign material were also injected into the bunny at regular intervals. After 6 weeks, the rabbit's blood serum contained high levels of a PG-specific antibody that was then purified away from the rest of its blood. Antibody so obtained was a very useful tool. As in an animal's immune system, the antibody would recognize and bind only to its specific antigen, in this case, PG protein.

Biochemists had been using antibodies to study specific plant proteins for many years. By grinding up tomatoes collected at various stages during the ripening process and adding PG-specific antibody to each different fruit sample, for example, the relative amount of PG protein present at each stage could be determined by comparing the amount of

* The abundance of PG protein in tomato fruit was a double-edged sword for the Flavr Savr tomato project. It was helpful because it made it relatively easy to produce PG antibody. But it was detrimental because the very high level of PG protein in ripe fruit also implied that the PG gene was turned on to a similarly high level, a level that might ultimately prove to be too overwhelming for the antisense "gene shut-off" strategy. And any amount of PG protein produced from an incompletely shut down gene might be enough to get the tomato-softening job done.

antibody associated with it. (Antibody bound to its antigen can be visualized by means of a dye attached to the antibody.) This was a very good trick, provided it was proteins that you wanted to study.

Molecular biologists, who study genes, as opposed to proteins, put a new twist on the old trick. They put foreign pieces of DNA, among them, they hope, genes or parts of genes they are interested in, into well-studied simple organisms that can be easily manipulated in the lab. Bacteria or viruses that attack bacteria (bacteriophages) serve this purpose well. Like a player piano slipped a new roll of music, each resulting "recombinant" organism will then "read" the foreign genetic information it has received as its own and produce the protein or part of the protein coded for in that foreign DNA. The bacteriophages producing PG protein are then identified using PG antibody.

Bill and Ray, as well as Alan and his team and Don Grierson and his colleagues, utilized this method to identify the PG gene. They inserted representations of all the genes that were turned on in ripe tomatoes (identified by the fact that they were undergoing transcription, i.e., producing RNA, in the fruit) into their lab organism (cloning vehicle) of choice. Each bacteriophage that successfully took up a tomato gene copy was then induced to multiply. It thereby cloned itself and the tomato gene sequences it carried along with it. PG antibody was then added to tens or hundreds of thousands of these recombinant organisms in order to identify individual ones that contained the PG gene copy. Success was largely a matter of numbers and visualized as little blue dots on round pieces of filter paper.

As it turned out, it took a little luck along with cutting-edge technology in order for Ray and Bill to identify their PG clone. I came across Bill Hiatt perched on the edge of the stool at his lab bench furiously flipping through one of his old lab notebooks several years after the PG gene had been obtained at Calgene. He was looking up dates and supporting materials for one or another of the patents on which he was an inventor.

"Look at this," he said as I walked by. "Wouldn't you think that it was this clone that contained the PG gene?" The blue dot he pointed to was definitely the darkest one on the coaster-sized piece of filter paper he had glued into his notebook. Most of the couple of hundred other dots on the same filter were gray, although a couple were light blue. Knowing that the method he had used to identify the correct clone was based on a blue indicator dye, I nodded in agreement.

"Actually, this was the right clone," he said, indicating a small dot, barely blue and barely contained on the edge of the filter. If the placement of the filter paper during the experiment had been slightly different, Bill and Ray would never have seen the clone. "The real clone nearly got away," he said with a flourish, a big grin on his face.

Once a particular PG clone was identified, several other pieces of information were used to verify that it actually contained a copy of the PG gene. Translating back from the language of proteins to the language of genes and RNA, the known PG protein code was compared to that predicted by the DNA code of a putative PG clone, for example. A major key to the success of this cloning strategy, however, was to obtain and characterize good, clean PG protein. For the protein part of the project, both Hiatt's and Bennett's groups collaborated with the same Australian scientist, Colin Brady. Brady was, in fact, made an author on the Calgene paper.[5]

Hiatt's and Bennett's groups were also similar in that they both received funding for their projects from big-name tomato processing companies. Bennett received "research gifts" from Chesebrough-Ponds and Beatrice/Hunt-Wessen Foods,[6] while the work at Calgene was sponsored by Campbell Soup.[7] As it became obvious that the two groups were duplicating each other's work, Al Stevens, the executive at Campbell Soup who spearheaded the contractual arrangement with Calgene, invited Alan Bennett to lunch with him and Bob Goodman, Calgene's VP of R&D. Al suggested collaboration. Alan was willing and offered a specific plan. But it was already too late for such a collegial arrangement. Like most university professors at the time, Alan hadn't filed a patent application to protect his work. Calgene, of course, had, and for "proprietary reasons" Bob Goodman therefore refused to collaborate. So instead of an efficiency operation, Al's lunch became a catalyst for animosity between the company, on the one hand, and the academic, on the other.

Industry vs. Ivory Towers

The issue of how science was conducted in an industrial, as opposed to an academic, setting was supersaturated with contention before that fateful interaction between Bob Goodman and Alan Bennett. It was generally believed, for example, that industrial scientists "had a competitive advantage over their university colleagues because they were working in large teams with considerable resources of space, equipment, and staff flexibly available for timely use on specific projects."[8] Scientists at U.C. Davis, as well as at other academic institutions around the country, harbored the impression that Calgene, just like other, much larger ag biotech companies, had "deep pockets" and therefore conducted "big science." Rumors spread that Calgene had put a huge team of scientists on the PG cloning project in an effort to beat out the U.C. Davis group. Alan himself declared that Calgene "outspent and outgunned" him.

In reality, except for the collaborative effort with Colin Brady, the PG gene was cloned at Calgene by a team of only two people, Ray Sheehy and Bill Hiatt. And, although you'd never know it from the persistent "deep pockets" rumors, Bennett soundly beat the Calgene group to publication on the PG cloning project. His paper, "Molecular Cloning of Tomato Fruit PG,"[9] was published in September 1986. Don Grierson and his group also beat Hiatt and Sheehy. They published their PG cloning paper in November of that same year. Bill and Ray's paper, "Molecular Characterization of Tomato Fruit Polygalacturonase,"[10] wasn't published until the following June. Deep pockets or no, the Calgene team had definitely lost the first battle of the war.

The race to demonstrate that tomato genes could be shut down using antisense technology was just as close as the race to clone the PG gene. Although Alan Bennett and his team did not enter the antisense race, Don Grierson and his group did, and, theoretically (based on their publication date), they had the advantage of obtaining their PG clone earlier than Ray and Bill did. This was the publication race that Bill was concerned about when the Flavr Savr tomato concept first hit the newspapers in August 1988.

The Race to "Antisense" the PG Gene

With their PG clone in hand, cutting the PG gene out of the bacteriophage DNA, flipping it around into the antisense orientation, and recloning it were relatively easy genetic engineering manipulations for Ray and Bill to make. What they needed help with was a way to get the new antisense-PG, or Flavr Savr, gene into the DNA of an entire mature tomato plant so that they could determine whether the new gene had any effect on the level of PG protein in its fruit. For this step in the Flavr Savr tomato project, working with tomato plants as an experimental system was a disadvantage. The model systems for these types of gene transfer, or transformation, experiments at the time were the noxious tobacco plant and the relatively frivolous petunia plant. Flower-bearing individual plants with new, genetically engineered genes in their DNA could be routinely produced in these two species. The same technology for producing transgenic tomato plants was not available, at least not in any reliable kind of way. However, JoAnne Fillatti and her coworkers at Calgene had been successful in their efforts to develop a routine method for transforming tomato plants with foreign genes. Her timing couldn't have been better, at least as far as Bill and Ray were concerned. (She was, however, beaten to publication herself by a scientist associated with Monsanto Co., Calgene's archrival in the ag biotech business.) JoAnne published her method[11] at about the same time that Ray and Bill published their PG gene-cloning paper. Ray shipped the Flavr Savr gene off to a member of JoAnne's group, Kristin Summerfelt, and waited for her to work the "magic" of producing tomato plants that carried the new gene in their DNA.

Plant Transformation Magic

The most commonly used "magic" involved in transforming plants takes advantage of a method of moving genes from one organism to another that is found in nature. Just as biochemists utilize the natural functioning of animal immune systems to create antibodies for research purposes, plant cell biologists utilize the natural transformation capabilities of a soil bacterium, *Agrobacterium tumefaciens*, to insert selected segments of DNA into tomato and other plants. This bacterium causes the formation of a gall on the crown of the plants it infects (crown gall disease) by inserting a segment of its own DNA into the DNA of a susceptible plant. By having its genes expressed in the infected plant, it forces the plant to produce substances the bacterium needs to grow. Genetic engineers manipulate this parasitic relationship for their own purposes by eliminating the bacterial genes in the segment of DNA that is transferred to the plant (the T-DNA). This manipulation effectively "disarms" the bacterium of its pathogenic capabilities—it can no longer cause galls on plants—and it also provides room in the T-DNA for a genetic engineer's interesting gene or genes, in this case, the Flavr Savr gene.

It was important, however, to make sure that every cell in a transformed plant contained the new gene. Therefore, individual cells from tomato leaves, stems, or other adult plant parts that could be coaxed into regenerating into full grown, fertile, cloned plants (in a manner not so different from the creation of the sheep Dolly[12]) were incubated with agrobacteria harboring the Flavr Savr gene in their T-DNA. The tomato and bacteria cells were provided a laboratory environment conducive to carrying out both the DNA transfer process and the regeneration of a cloned tomato plant. The two organisms were incubated together under specific conditions of light, heat, and supporting media, including plant growth hormones in proper proportions. Tomato cells that had permanently acquired the T-DNA were then identified by means of a selectable marker gene included in the T-DNA along with the Flavr Savr gene. The selectable marker gene coded for a protein that provided cells expressing it resistance to an antibiotic.

By including that antibiotic in the medium used to induce tomato cells to multiply and develop into full-grown tomato plants, only the tomato cells expressing the selectable marker gene (and therefore containing that gene and the Flavr Savr gene adjacent to it) would survive to go through the regeneration process. The antibiotic was used to select the transformed cells from among hundreds or thousands of nontransformed cells. Like the sheep-cloning experiment,[13] transforming tomato plants was a very inefficient process. The use of a selectable marker gene in this manner increased efficiency dramatically. But most other ag biotech companies used the same one, and there were fears that allowing it into tomato and essentially every other genetically engineered crop on the planet could somehow compromise treatment of humans with antibiotic drugs. For this reason, Calgene's selectable marker gene would take on a life of its own later, when it came time to get regulatory approval from the U.S. Food and Drug Administration (FDA; see Chap. 3).

It took about 6 months of transformation-regeneration magic before Ray and Bill had mature tomato plants, transformed with the Flavr Savr gene, that were producing fruit. In order to speed up the analysis of those tomatoes, Bill hired Matthew Kramer in 1987. Matt received his B.S. in microbiology and his M.S. in agronomy at Montana State University. He met Bill Hiatt at a scientific conference where they'd been stationed next to one another during a poster session. They got to talking, and Bill asked Matt to come work for him at Calgene. Matt liked Bill and the idea of working for a company and promptly left the Salk Institute in San Diego for Davis. Matt was given the title Product Development Specialist. Calgene was preparing for product mode.

Matt was especially nervous his first couple of months on the job, not because of his work responsibilities, but because his wife was hospitalized with a serious condition soon after they arrived in Davis. "I almost lost her," Matt told me later. His nervousness made him a bit of a klutz in the lab. At a rate of about one every other day, he knocked a full box of sterile pipette tips onto the floor. In addition to the waste involved, each of these mishaps made a mess analogous to that created by dropping a full box of brightly colored, bouncing toothpicks. The last thing a scientist wanted was to be saddled with a "bad hands" label. Not surprisingly, Matt's case of bad hands cleared up as soon as his wife recovered.

Matt was a company man. Unlike Ray, Matt loved to give presentations and send memos. Like Ray, he wasn't always sure how to respond to Bill Hiatt's humor. On Matt's first day of work at Calgene, for example, he entered the cavernous warehouse-type structure housing what Calgeners referred to as the "main lab" and saw Bill at his bench on the other side of the room. Bill looked up as Matt approached him, said, "Just a minute, I'll be right with you," and lifted a liter flask full of growing bacteria in each hand. He inhaled deeply from the open spout of each flask before looking back at Matt. "Ah, now then, let's talk," he said. It was the start of a beautiful relationship.

Unlike Ray, however, Matt didn't give Bill much reason to tease him. Matt supported Bill's ideas in meetings and seminars and yet wasn't afraid to disagree with him on important issues. He provided Bill with accurate information from scientific conferences so Bill didn't have to attend them. Most important, Matt supplied Bill with what Bill felt were the key data demonstrating that antisense technology could effectively reduce PG protein levels in tomato.

Antisense Works

Prior to Matt's arrival at Calgene, Ray had shown, in his quick and dirty way, that fruit from the first group of Flavr Savr tomato plants had 63 to 83 percent less PG protein activity (i.e., ability to break down pectin) than did fruit from wild-type tomato fruit. Ray had also determined that the Flavr Savr gene was turned on—that is, RNA transcribed from it was present—in ten different transformed plants. However, he had not attempted to quantify how well the Flavr Savr gene was working in each plant.

Matt took nearly the opposite approach. Instead of looking at groups of transformed Flavr Savr tomato plants, he meticulously characterized just one. He carried out the same types of experiments as Ray had, but he took extra steps so that he could draw conclusions about just how much RNA had been produced by the Flavr Savr gene compared to the resident PG gene. And he amassed detailed information on wild-type tomatoes so that he had plenty of control data. He also examined just what happened to the expression of both the Flavr Savr and the resident PG gene throughout the process of fruit ripening, not just in fully ripe fruit. The latter data added a flair of basic research to a story that could otherwise be criticized, at least in prestigious scientific journals, as applied science.

Ray's data were perfectly valid, just not nearly as "pretty" as Matt's were. In some ways, Ray's data were more valid. Every transformed plant turns on its inserted gene to a different level, and, therefore, many independently transformed plants should be analyzed in order to get a good idea of the big picture. An anonymous peer reviewer of Ray and Matt's antisense paper, in fact, criticized it for confining so much of the analysis to the one plant that Matt had studied so well.

But Bill liked clean, precise experiments. Bill liked numbers. And Matt's data indicated that only 10 percent of the normal amount of PG RNA was present in that particular transformed plant. The PG gene had been effectively shut down by 90 percent! That 90 percent reduction in PG gene expression was something Bill and Calgene management could, and did, hang their hats on. Bill gave Matt credit for that. In fact, Bill said, "If it hadn't been for Matt Kramer, our antisense paper would never have been published at all." Matt had assumed the role of Bill's right-hand man.

A Second Race is Lost

But, despite Matt's and Ray's and Bill's efforts, the Calgene team was beaten to the punch again by Don Grierson's tomato group. (Not that this loss necessarily reflected badly on the scientific skills of the Calgene team, in a man-to-man kind of way at least. There were *seven* authors listed on the British group's paper!) Bill's fears of being scooped had been realized. Grierson's manuscript, which documented a shutdown of the PG gene in tomato similar to that which Matt, Ray, and Bill had found, appeared in the August 25, 1988, issue of the journal *Nature*,[14] just 10 days after Calgene's first Flavr Savr tomato press release. The antisense paper by Sheehy, Kramer, and Hiatt didn't appear in print until the December 1988 issue of the *Proceedings of the National Academy of Sciences*.[15]

These two papers were among the first transgenic studies in tomato plants ever published. What's more, most of the transgenic plants engineered before then—and there were a handful of them—had been genetically modified to tolerate some type of herbicide. As opposed to herbicide tolerance, an agronomic trait of more interest to farmers than to consumers, shutting down PG in tomatoes hinted at a higher-quality commercial product that consumers could relate to. Therefore (as described in Chap. 1), Calgene management made sure that the results (and then some) of the Calgene study were splashed all over the newspapers. But these were scientific consolation prizes. They did not reduce the sting of losing a second scientific battle in the Flavr Savr tomato war.

The Race with Meaning for the "Real" World

Contrary to the optimistic implications in the popular press during August 1988, neither the Calgene group nor the group from the University of Nottingham presented any data in their scientific papers on the effect of successful PG protein reduction on tomato fruit softening. Instead, both teams focused on the successful use of the antisense technology to dramatically shut down, although not shut off, the PG gene and skirted the issue of what difference that shutdown made in the real world of tomatoes. Hiatt and coworkers were particularly elusive on the subject of fruit softening, stating only that the "ability to regulate the expression of the PG gene has agricultural importance because of the role of the enzyme in fruit softening."[16] Their reluctance to go out on a limb was largely due to Bill's firm belief in "letting the data speak for themselves." And at that point there were simply no data on tomato firmness. Sheehy et al. did add, however, that they were examining "additional physiological parameters of fruit development related to PG activity,"[17] meaning whether the fruit was firmer.

Grierson and coworkers were somewhat more, although still cautiously, gregarious. They went beyond general statements of the historical correlation between PG protein and tomato softening, although they made one of those too, and actually tested the waters of applied science. Of course, that's just where they found controversy lurking. For example, "fruit shelf-life and processing quality,"[18] traits related more to traditional gassed green and canned tomatoes than to vine-ripened fruit, were the potential applications for a tomato business they mentioned. But even more controversial was the statement, made without any supporting data, that "attempts to measure softening by compressibility show no differences"[19] between the genetically engineered tomatoes and the control fruit.

We at Calgene were, understandably, disappointed that Grierson's group had not detected any improvements in tomato firmness in their antisense-PG tomatoes. But we noted that the jury was still out. We believed that Grierson and company regarded their results as preliminary, since they had gone on to say that "further work is in progress to define softening and the role of PG."[20] (The fact that the "attempts to measure softening" statement had been added to the manuscript after Alan Bennett had peer-reviewed it for *Nature* served as additional evidence of the preliminary nature of the results.)

The Squashing Technique for Measuring Tomato Firmness

Tomato firmness measurements, and lots of them, were the only way to settle the softening issue. But the methods of measuring tomato firmness at the time didn't instill much confidence even in those of us who made them. During a seminar in 1993, Don Grierson described the two most common methods as the "drop a heavy weight on the fruit and measure the degree to which it is squashed" method and the "determine how hard you have to poke it with a pencil before it gives" method.[21] Obviously, these techniques left much to be desired. The crude "apparatus" used at Calgene to measure tomato softness utilized the squash (compressibility) technique. Bill Hiatt hated it. At one point he ordered, "I don't want any more measurements made with that thing. Find some other way to determine fruit firmness."

The offending "device" consisted of a mangled lead weight, crudely horseshoe-shaped, with a mass of 500 grams, that was precariously placed around a steel bar set against the top of a fruit (very carefully so that the weight would not fall off). The steel bar was connected to a dial, and as the fruit was compressed the dial would spin. The amount of spinning indicated, in some arbitrary units, the degree of squashing (movement of the bar) that had occurred. Recording these compressibility measurements required the attentive interaction of two humans. One noted the number of spins that the dial had taken while the other noted the time during which said spinning occurred.

Bill felt that the fact that such a device was used at all in a company like Calgene, theoretically "on the cutting edge of technology," was bad enough. But to use it to measure the trait so critical to the success of what was sure to be the company's first genetically engineered product was almost more than he could stand. He had some consolation in the fact that the competition in Great Britain used a modification of the same device, although in their version the dial spins per unit of time were automatically measured and recorded by a computer. They, it appeared, had at least entered the twentieth century of squash technology. A relatively minimal investment of time—associated, like almost everything else during that era in Calgene history (see Chap. 5), with a somewhat substantial investment of money—was spent investigating various extremely high-tech, yet unproven, alternatives. In the end, the low-tech squashing technique of measuring tomato softness prevailed at Calgene.

The Opposite Tack:
Adding PG to Tomatoes

Ironically, the first hard data (no pun intended) on the role of PG protein in tomato softening, at least as demonstrated in genetically engineered plants, was published by Alan Bennett and his colleagues a mere month after the Calgene antisense PG paper appeared in print. Since it had been clear from the lunch with Al Stevens that collaborating with Calgene to antisense PG was impossible and competing directly with the company wasn't appealing, Alan had taken a different approach to studying the role of PG protein in tomato fruit. He inserted his PG gene clone, in the correct "sense" orientation, into a tomato variety that was unable to produce ripe fruit. This tomato variety, containing a mutation called *rin* (for *r*ipening *in*hibitor), produced fruit that lacked PG protein and failed to soften. Because *rin* tomatoes also failed to turn red properly and did not acquire ripe tomato flavor, the theory was that the mutation causing these multiple defects must have occurred in a single gene that controlled the function of many other genes, one of which was the PG gene. By inserting a viable PG gene, via genetic engineering, into this mutant tomato plant that lacked a functional PG gene of its own, Alan and his colleagues tested whether production of PG protein alone was sufficient to cause the mutant tomato fruit to soften. Their conclusions were flaunted in the title of their paper, "Expression of a Chimeric Polygalacturonase Gene in

Transgenic *rin* Tomato Fruit Results in Polyuronide Degradation but *Not Fruit Softening* [emphasis added]."[22] Expression of the inserted PG gene led to production of PG protein in the fruit of the mutant plant, although to significantly lower levels than those found in wild-type fruit. The PG protein produced was capable of breaking down pectin in a test tube and was therefore active. However, the active PG protein did not make the mutant fruit any softer, as determined using the squashing method. Alan Bennett's conclusion was that active PG protein on its own was not sufficient to carry out fruit softening. The implication for the Flavr Savr tomato was that PG might not be the major player that Calgene was banking on in the tomato-softening game. Even if PG could be eliminated in vine-ripened tomatoes, other factors that contributed to tomato ripening might still soften the fruit.

Bennett and coauthors also paraphrased Grierson's statement in his 1988 antisense PG paper[23] that "attempts to measure softening by compressibility show no differences" as additional support for their results. Calgeners thought citing the Nottingham group's undocumented "attempts" was hitting their Flavr Savr tomato project below the belt. And every time Alan projected that isolated statement onto the big screen at scientific conferences during the several years between the publication of Don Grierson's initial paper and that of his subsequent manuscripts, which actually contained compressibility data, Bill Hiatt and his colleagues felt they'd been suckerpunched. As a result, although Bill and Alan remained cordial face to face, chatting at their kids' soccer games, theirs became an intensely adversarial relationship. Straying from the confines of hard scientific data had stirred up a hornet's nest around the Flavr Savr tomato.

Eliminating PG:
The First Firmness Results in Antisense Plants

Don Grierson and seven colleagues, in a collaborative effort with ICI Seeds, were next to publish a paper containing firmness measurements.[24] It had been nearly 3 years since the first antisense PG paper his group had published, and they now had hard data on the tomato firmness issue. They had performed huge compressibility experiments and drawn statistically significant conclusions from them. Their interpretation of their data was straightforward and, unfortunately for Calgene, the same as the preliminary conclusion they had drawn 3 years earlier: firmness in tomatoes genetically engineered with their version of the Flavr Savr gene was no different than firmness in control fruit. The Flavr Savr tomato hypothesis did not seem to be panning out.

Ironically, Bill Hiatt and his team at Calgene provided their own fuel for the anti–Flavr Savr tomato fire in a paper they published in 1990 in a book edited by Alan Bennett. In retrospect, it's amazing that Calgene management allowed Bill to publish it.

The paper, "Field Evaluation of Tomatoes with Reduced Polygalacturonase by Antisense RNA,"[25] was a report on Calgene's first field trial of Flavr Savr tomatoes. The trial was planted on November 31, 1988, in Guasave, Sinaloa, Mexico. Conducting their maiden field trial of genetically engineered tomatoes in a foreign country, and a developing one at that, was not necessarily a bright public relations move for Calgene. But the choice had been to plant in Mexico immediately or wait to plant in California until the following spring, and "wait" was quickly being eliminated from the Calgene vocabulary.

The results from the trial were mixed at best. There was statistically significant improvement in the consistency and viscosity of processed tomato juice, for example. There was hope, therefore, that tomato processors might be able to minimize their "break" step, the treatment used to destroy the activity of PG protein in harvested tomatoes, and consequently their energy bills without running the risk of producing runny ketchup. But the improvement was observed in only 20 percent of the plants transformed with the Flavr Savr gene. And the extent of the improvement, while statistically significant, was not great. In fact, Matt concluded that "commercial application of lowered break temperatures may require elimination of additional PG activity."[26]

Although solving the PG problem for processing tomatoes was, to a large extent, just a matter of degree (the more PG that could be eliminated, the more money processing companies could save), the double-edged PG sword still seemed poised to smite the project. Perhaps there was just too much PG protein in tomatoes to be able to reduce it enough using antisense technology to make any real difference. Or maybe Alan Bennett was right, and another enzyme or enzymes involved in tomato softening would also need to be eradicated. On the other hand, maybe Matt and his colleagues just had to try harder. In any event, lowering break temperatures was a concern with processing tomatoes, not fresh-market fruit, and was therefore not Calgene's top priority.

Disappointing News for Fresh-Market Tomatoes

More discouraging for fresh-market tomatoes was Matt's description of a trait called *field holding ability*. Flavr Savr tomatoes produced the same total yield of fruit as did control plants, but they had significantly less rotten fruit than control plants at harvest time and beyond. The yield of usable, nonrotten ripe fruit was therefore significantly higher for the Flavr Savr tomato plants.[27] This was true, Matt noted, not because they were maturing later than control plants, however. The Flavr Savr tomatoes grew normally in the field, they flowered on time, and they produced fruit that ripened similarly to control fruit, "including the initial phase of fruit softening."[28] Because they ripened normally, Matt concluded that they remain "intact in the field for a longer period of time" because the "predominant role for the major portion of PG activity is a deterioration of fruit integrity during the overripe stage" of tomato ripening.[29]

This field holding ability was great news for tomato processors who pick their tomatoes ripe; they could wait for the entire field to ripen, pick once, and not have to worry about rotten fruit. But if Flavr Savr tomatoes went through the initial phase of fruit softening normally, where was the advantage for a fresh-market tomato business? The whole idea was that Flavr Savr tomatoes would be firmer as they ripened, not just as they rotted, so that they could be left to ripen on the vine longer and therefore taste better but still survive being trucked to grocery stores. If the Flavr Savr gene improved tomato firmness only during the overripe, or rotting, phase of tomato ripening, it was not going to help Calgene or any other company sell vine-ripened tomatoes. Tomato processors accept, even expect, that a certain part of their tomato harvest, at least the part on the bottom of the truck beds, will arrive flattened. Grocery store produce managers don't have the same expectations. Clearly, this was not the answer Calgene had been looking for.

The Patent Interference

During those dark days when proof of the Flavr Savr tomato concept was not just nonexistent but, in fact, distinctly negative, Bill Hiatt got a bit of good news. U.S. Patent No. 4,801,540, "PG Gene and Its Use in Plants," was issued to Bill and Ray and three additional Calgene colleagues on January 31, 1989. But before Bill had time to celebrate, his patent was challenged by Don Grierson and his coinventors on behalf of ICI Seeds. Through files related to that challenge, Bill learned just how close the race to patent the PG gene had been.

Based on U.S. patent laws then in effect, Grierson's original invention date related not to his research activities in Great Britain but to the date he had established an appropriate disclosure of that invention, in the form of a scientific manuscript, on U.S. soil. Assigning a date for Grierson's PG cloning paper was less straightforward than it might have been, however, because it was first necessary to establish whether the clone was in fact a PG clone. Grierson's group originally claimed in a 1985 publication to have cloned PG.[30] But, it turned out, what they had identified instead was a clone for some other tomato gene. Alan Bennett and his coworkers politely pointed out this fact in their September 1986 paper by stating that Grierson's previous work "may be incorrect."[31] In November 1986, Grierson's group reported the DNA sequence of a bona fide PG clone and also conceded that the clone they had analyzed in the 1985 publication did "not code for PG."[32] The date of that 1986 publication, November 11, was the date that had been used in determining which group should be granted the PG patent. Bill Hiatt and his Calgene colleagues had filed their PG data with the Patent Office less than a month earlier, on October 17.

That was far from the end of the PG patent saga, however. Unlike countries that grant patents on a "first to file" basis, the U.S. Patent Office awards patents based on who can prove they were "first to invent." In their challenge to Calgene's patent, Grierson and company claimed that they had presented their "invention," including the DNA sequence of their PG gene clone, at a scientific conference held on U.S. soil some 3 months earlier than the November publication date of their PG cloning paper, and therefore 2 months earlier than Calgene had filed its PG sequence data. This put the two groups in a situation called "patent interference." As a result, the Patent Office ignored their original filing dates and, in an effort to be fair to the party that was truly first to come up with the idea, set out to determine the actual date of invention.

Since the August 1986 meeting, the Molecular Biology of Tomato, had been held at U.C. Davis, several Calgene scientists (although not Bill Hiatt because he had been ill) had heard Grierson's seminar. Not one of them remembered seeing a PG clone sequence despite the fact that it would have been the highlight of the meeting for them. Unfortunately, not one of them possessed any notes that could have proved or disproved Grierson's claim. But, so it seemed, neither did he. What he did have was a sworn statement from Alan Bennett, who had also attended the conference, supporting Grierson's position in the patent interference. We, at Calgene, simply couldn't believe Alan had seen a PG clone sequence during Don's talk when we hadn't, and Roger Salquist could contain himself no longer. Believing that "any press is good press," Roger referred to the English group as "those Bozos from Britain" in an oft-quoted newspaper interview. (I saw Don Grierson at a conference several months later. He asked me to "tell Bill Bozo says hi.")

In response to the PG patent interference situation, Bill Hiatt's and Ray Sheehy's laboratory notebooks were thoroughly examined. As was done for all lab work at Calgene, Ray and Bill kept experimental notebooks to document their experimental work. Academic scientists use lab notebooks, too, but the notebook pages of industrial scientists make particularly good patent documents. Every page in a Calgene scientist's notebook was signed and dated everyday by the scientist conducting the experiments. They were also signed and dated by another scientist, who served as a disinterested "witness" of the experiments. Ray's and Bill's notebooks stood up to the scrutiny. Although they may have contained some examples of what not to include in a lab write-up (the rest of us were reminded not to use them as personal journals, for example), they established exactly when Ray and Bill had in fact obtained their copy of the PG gene.

When all was said and done, Bill and Ray retained their issued patent on the PG gene sequence and its use in plants.[33] Now all they had to do was prove that using it to get rid of PG protein in tomatoes was of value to a fresh-market tomato business.

The Squashing Technique Pans Out

Bill Hiatt didn't dwell on the results of the first Flavr Savr tomato field trial. While he and Matt Kramer had stated in the 1990 publication that Flavr Savr tomatoes went through the initial stages of fruit softening normally, that paper, like Don Grierson's 1988 manuscript before it, contained no actual compressibility measurements. Why they had published such a detrimental statement at all, especially with no supporting data, is incomprehensible. In order to fully understand the effect the Flavr Savr gene did or did not have on tomato softening, compressibility measurements, and lots of them, were needed. Bill assigned Matt Kramer and the newest member of Calgene's Flavr Savr tomato team, Rick Sanders, that awesome task.

Unlike Ray and Matt, who had been brought on board as new employees at Calgene, Rick had been reassigned to the Flavr Savr tomato team from within the company. A mere 6 months after I'd been hired and in the midst of the 1988 Flavr Savr tomato PR frenzy, Bill came into my office and shut the door. I didn't realize it at the time, but closed doors at Calgene were a pretty reliable sign that a lay-off was imminent. "I have to make a decision about letting an employee go," he said as he sat down. "It comes down to a choice between [a man who had been working on my project, an effort to increase "solids" in tomatoes] and Rick Sanders. I'd like your input."

I was not at all pleased at being placed in such a position. This was not the introduction to industrial science that I'd been hoping for. It was, however, commendable of Bill to solicit my opinion, since the decision had the potential to affect my workload and scientific strategy. On the other hand, these types of decisions wreaked havoc on the personal, as well as the professional, lives of the people let go. Bill, a senior manager at the time, was paid to make those kinds of calls. I wasn't.

Rick A. Sanders received his B.A. degree in biology from California State University, Chico, in 1983. He came to work at Calgene in 1984, originally as a dishwasher. He quickly worked his way up the company ladder, though, to research assistant. In August 1988, Rick was working with a group of scientists at Calgene trying to transform and regenerate corn plants, and the entire corn program was about to be terminated. Corn plants, at the time, were nearly as hard to clone as sheep. Bill, who saw Rick socially, duck hunting, could save Rick's job by transferring him into the tomato program.

But I didn't know any of this at the time. I didn't even know who Rick Sanders was, and I told Bill so. I also told him what little I knew of the research associate who was working on my project. He left my office disappointed that I couldn't "weigh" the two candidates against each other and, I believe, still undecided.

A few days later I learned that Bill had chosen Rick. Rick was made a product development specialist and assigned to work for Matt Kramer conducting tomato field trials. I was disappointed that Rick wasn't assigned to work with me. (After a couple more layoffs, I more readily accepted the increased work burden that always accompanied them.) Then and many times afterward, Bill said, "Bringing Rick on board the Flavr Savr tomato project was the best decision I've ever made." I had to agree with him.

In addition to the love of the (duck) hunt, Bill and Rick shared that love of professional sports, especially American football, that borders on the fanatic. They had different favorite teams. Rick grew up in northern California and was therefore a life-long San Francisco Forty-niners fan. Bill had adopted the then Los Angeles Rams during his postdoctoral years in southern California. But rooting for different teams only served as fuel for their fiery discussions on Monday and Tuesday mornings during football season. Rick's ability to talk professional sports, in addition to duck hunting, bonded him to Bill in a way that was not possible for Ray or Matt.

That is not to say that Ray and Matt didn't like sports. They were both quite athletic, in fact. Ray coached, managed, and played for Calgene's long-standing softball team. Matt loved the great outdoors and was especially skilled at cross-country skiing, particularly down "hill." He and his wife once took my husband and me skiing on Mt. Shasta, the 14,162-ft–high volcano in northern California. We had the advantage on the aerobic uphill climb. But on the downhill, Matt and Kathy left us in the powder with their beautiful, effortless-looking telemark turns. However, neither Matt nor Ray was your typical armchair athlete, watching the games, reliving them through newspaper articles, and recounting them with fellow armchair athletes. Bill and Rick were. (They were "real" athletes as well. Bill played racquetball and Rick ran and played alongside me on Calgene's basketball team, the Biohazards.)

Rick also got along well with his immediate boss, Matt Kramer, as well as with everyone else at Calgene, for that matter. He was pleasant, friendly, thoughtful, and, most important, he had a work ethic that rivaled Matt's. The two men made a great team. In slides shown during company meetings, one of them published in the October 1, 1990, issue of *Time* magazine, the two of them are seen standing next to piles and piles, hundreds, maybe thousands, of tomatoes from a field trial they'd conducted in Florida and on every one of which they had taken firmness measurements.

All of their hard work paid off. They reported, in a manuscript published in 1992,[34] that Flavr Savr tomatoes picked after they had turned red on the vine "were as firm as their nontransformed counterparts harvested green and treated with ethylene." They had also sorted the fruit "over a commercial packing line to simulate handling and storage conditions relevant to industry practices" and felt comfortable enough to predict that Flavr Savr tomatoes "may benefit commercial producers of fresh market tomatoes . . . by allowing for the harvest of a greater proportion of fruit with color while avoiding the usual losses attributed to the packing and shipping of vine-ripe fruit." This was the missing piece of scientific evidence, the support for the dream of a fresh-market tomato business that Bill Hiatt and Calgene management had been waiting for. They took it and ran.

The Competition Concedes

In the end, Don Grierson backed down on his claims that fruit firmness was not altered in Flavr Savr tomatoes. In a seminar he gave in 1992, he made a point of addressing the by then quite controversial issue. Before his formal presentation, the keynote address of an Antisense Symposium at Iowa State,[35] he spoke to the audience. "I have been told that people are saying we see no improvement in the firmness of our genetically engineered tomatoes. I'd like to clarify the situation." He went on to say that, while they had found no changes in softening in their initial experiments, since then, at least in some tomato varieties, they had found fruit significantly firmer due to expression of their version of the Flavr Savr gene. He stressed that, in their hands, the type of tomato into which the gene was inserted and the extent to which that tomato would have softened normally were the factors that made the difference between the two sets of experimental results. He was more succinct in print. In reference to a 1993 publication,[36] he simply stated that "small but statistically significant differences in fruit firmness were measured in fully ripe fruit."[37]

Alan Bennett, too, was a party to a kind of scientific retraction of his claim that PG protein had less to do with tomato softening than previously thought. In a collaborative study carried out among several academic labs and with which Alan Bennett and I were both associated, the very same transformed *rin* plants that had been analyzed for Bennett's 1989 publication were reexamined. In the earlier paper, Alan had left open the possibility that the levels of PG protein produced in the transformed *rin* tomatoes might be "insufficient," compared to levels found in normal nonmutant plant fruit, to cause fruit softening. Sure enough, the new data, reported at a scientific meeting at Fallen Leaf Lake, California, in the fall of 1992,[38] bore that possibility out. When the cloned PG gene in those plants was forced on to a higher level (through environmental manipulation or a greater number of gene copies), more PG protein was produced in the fruit, and, as a result, the fruit did in fact soften significantly. I couldn't help but ask the graduate student, Ron Martin, who reported the findings at the meeting, "Don't your results indicate that PG [protein] is not only necessary but also sufficient to cause tomato softening?" The academic audience laughed nervously, obviously aware of the ongoing controversy. Alan was not present to support or refute my comment, and Ron graciously sidestepped the issue, offering that his work did not rule out, in fact supported the idea, that other, as yet unidentified, factors also contributed to tomato softening. Nevertheless, Ron concluded from his studies that PG protein played a major role in the softening process.*

Scientific Vindication

So, in 1992, Bill Hiatt not only published his own evidence that the Flavr Savr tomato could benefit a fresh-market tomato business, but he also learned that both of his adver-

---

* At this writing, nearly 8 years later, interpretation of Ron's data remains controversial. His results still have not been published in a peer-reviewed journal.

sarial competitors had scientific results that put the project in a more favorable light. He had been scientifically vindicated. Nineteen ninety-two was a banner year for Bill in another way. Despite repeatedly losing out to Grierson and Bennett in the race to publish, in the end Hiatt and his coworkers claimed the real prize, at least as related to building a business. In April 1992, Calgene was issued a U.S. patent that covered the use of antisense technology in all plant species.[39] The concept behind the Flavr Savr tomato appeared not only proven but also protected.

The stage was now set. With patent protection in place and some scientific evidence to back it up, the Flavr Savr tomato was ready to have a business built around it. There was, however, one rather large piece of the puzzle still missing. As Bill Hiatt so eloquently put it at the time, "We have a potential product here, but we can't [legally] grow it, ship it or eat it." In order to handle this genetically engineered tomato like any other tomato and be able to sell it for human (or animal, for that matter) consumption, Calgene sought regulatory approval from the U.S. Department of Agriculture and the FDA. Gaining those regulatory approvals turned out to be quite a labor of love.

CHAPTER THREE

Virgin Territory

Plant Genetic

Engineering and

U.S. Regulatory Agencies

In August 1990, the members of Calgene's science staff who worked on tomato-related projects and those who worked on cotton had a joint meeting. Don Emlay, Calgene's director of regulatory affairs, had called us together. I, for one, entered that conference room feeling more than a little apprehensive.

Part of my apprehension had to do with our meeting facilitator. Calgene scientists hadn't seen much of Don, much less had any meetings with him, in the 3 years that he'd been with the company. The one time he had come out of the executive woodwork, however, in 1988, he had axed a major tomato project. The project under fire was a tomato version of BromoTol cotton. Don did not believe that Bro-moTol tomatoes would serve as a good first product for Calgene (or the ag biotech industry as a whole, for that matter), at least not the first time out of the blocks with a genetically engineered whole food. In contrast to the situation with cotton, in BromoTol tomatoes the bacterial protein that conferred tolerance to the herbicide bromoxynil was in a fruit meant to be eaten; herbicide residues would be an issue. And, based on the reaction of some consumers more than 10 years later to eating pesticide-producing, genetically engineered "Bt" (*Bacillus thuringiensis*) potatoes,[1] he was probably right. However, his announcement was not well received by Calgene's science staff, especially those who had had the tomato project pulled out from under them. And those of us who weren't directly affected by Don's project-cutting authority wondered what else he, as director of regulatory affairs, had been doing in a company that was still years away from having a product to be regulated. As we sat in Calgene's conference room A that August day, we were about to find out.

Don had been "rapidly developing an appreciation for the regulatory process required to obtain product approval."[2] He had joined the International Food Biotechnology Council, made up of some 30 food processors and biotechnology companies for the purpose of establishing scientific guidelines to assure safe production of biotechnology food products and pave the way for public acceptance of them. He had initiated discussions with the FDA on the safety of Flavr Savr tomatoes on February 2, 1989, only about 6 months after the original tomato shelf-life results.* Fired up by that first meeting, Don originally planned to submit a document asking the FDA for "approval" of the

---

* Although this seemed awfully early to those of us at the meeting who knew we were far from having a product to bring before the FDA (in the end, it would be nearly 4 years before our tomato was ready for market), Monsanto Corporation and Pioneer Hi-Bred International also had already met with the FDA.

Flavr Savr tomato by July of that same year, a mere 4 months later. But reality had prevailed, and he had instead spent the following 18 months convening various expert panels for advice, hiring a consulting firm—ENVIRON International Corporation—that specialized in getting drugs and foods approved by the FDA, and taking on a second-in-command, Keith Redenbaugh.

Keith, a Ph.D.-level cell biologist, had come to Calgene by way of the company's purchase, in June 1989, of its next-door neighbor, another ag biotech company, called Plant Genetics, Inc. He had been promised a research effort to direct at Calgene and therefore decided to forgo another job offer he had prior to the acquisition. Only a few months post-merger, however, he was unceremoniously relieved of his research-related duties by V.P. of Research and Development Bob Goodman and assigned to work for Don. Understandably unhappy, Keith spent the next 9 months or so biding his time, looking at his options. Despite (or perhaps because of) his lack of enthusiasm, Roger Salquist had given Keith a 20 percent pay hike. By August 1990, Keith was firmly on board at Calgene, and the company's general strategy for gaining official commercial blessings from the various U.S. agencies that oversaw regulation of genetically modified organisms had been formulated.

Under the Federal Food, Drug, and Cosmetics Act, the FDA was the agency with the authority to remove from the marketplace foods it considered unsafe and to hold the producers of those foods responsible for the safety and quality of the foods they marketed. At Calgene, obtaining FDA approval was considered imperative, for public relations reasons at least as much as for legal ones. The previous year, a batch of the food supplement L-tryptophan that had been isolated from genetically engineered bacteria was implicated as the source of an outbreak of eosinophilia myalgia syndrome (EMS) that killed 37 people and permanently disabled approximately 1500 more.[3] Although it was never established that the toxicity of the L-tryptophan was connected in any way to the fact that it was a product of genetic engineering, Calgene management worried that the question lingered among other Jurassic Park–inspired doubts in the collective public conscience. Especially because the Flavr Savr tomato was expected to be the first genetically engineered whole food to enter the marketplace, proceeding conservatively yet quickly to gain approval from the FDA was perceived to be the best approach.

The first item on Calgene's regulatory agenda was not to convince the FDA that its first potential product, the tomato, was safe. Instead, the plan called for demonstrating that the company's selectable marker gene, a gene conferring antibiotic resistance that was inserted into every one of Calgene's genetically engineered products, was both necessary to the process of producing transgenic plants and safe to use and consume.

Making FDA approval for Calgene's selectable marker gene the first goal of Calgene's regulatory strategy was key for several reasons. In 1990, Calgene didn't have any genetically engineered plants ready to go to market. In some cases the company's scientists had only good ideas for what they thought would eventually become genetically engineered products. Our putative first product, the Flavr Savr tomato, still required all kinds of product development, much more than anyone at Calgene dare to dream, before it would be a viably marketable tomato. Even the design of the Flavr Savr gene construction was still in transition. And, since employees at Calgene knew much more about designing genes than they did about tomato breeding, farming, handling, shipping, marketing, and so on, it was going to be quite a while before the Flavr Savr tomato was ready for market. The selectable marker gene, on the other hand, was ready to go. Calgene could get an early start, albeit a hypothetical one, to what was bound to be a long process with U.S. regulators by setting the ag biotech stage with its selectable marker gene.

The same selectable marker gene was used to transform all three of Calgene's core crops: tomato, cotton, and canola. If blanket approval of that one gene could be obtained from the FDA for all three crops prior to product availability, subsequent approval for the products themselves would be streamlined, involving only safety evaluations of the inserted gene or genes specific to that particular product. (Because of the fact that the Flavr Savr tomato and BromoTol cotton were so much closer to market than were any of the potential products being developed in Calgene's canola group and as a result of political maneuverings at Calgene, the selectable marker regulatory "blanket" was assembled without the participation of the company's canola scientists.) And, assuming that a Flavr Savr tomato–specific safety package could subsequently be put together and submitted to the FDA in a timely fashion (see Chap. 4), FDA review of the two petitions might then proceed in parallel. With a little luck, government approval and Calgene product development might come about simultaneously, thereby allowing product introduction to occur as quickly as possible.

By the summer of 1990, the staff at ENVIRON had put together a draft document designed to demonstrate to the FDA the safety of Calgene's selectable marker gene. It had been reviewed by one of Don's expert panels in July. Among other things, those experts found ENVIRON's draft too descriptive. It was time to beef up the "armchair" science contained in that document with data. It was time to bring in the scientists.

But we scientists were still reeling from the fact that, just weeks earlier, near the end of fiscal year 1990, the company had taken "significant steps to increase the focus of . . . R&D efforts on . . . core programs . . . and to reduce our total R&D expense to a sustainable level."[4] The "reduction" had, of course, taken the form of layoffs and, as usual, those layoffs had been primarily among the science staff. Those of us who remained were, as usual (it was already the third such reduction I'd survived in 30 months with the company), demoralized. The "focus," in which various scientific projects had been eliminated and the scientists associated with them (who remained employed at all) reassigned to other projects, had additional disheartening effects. Several such reshuffled scientists had been called to this first meeting of Calgene's regulatory science team.

Another reason Calgene scientists weren't very excited about tackling "regulatory science" that August had to do with the scientific culture at Calgene. An unwritten, unspoken (at least in public) hierarchy existed among the various scientific disciplines represented in the company. Cutting edge molecular biology was the science held in highest esteem. I was never sure just why this was so. The biochemistry and cell biology at Calgene, for example, were topnotch and equally important to the company's genetically engineered products. I suspect it had to do with the strong personalities of Calgene's highest-ranking scientists, molecular biologists and all men, and with history. The first tangible valuables the company possessed, after all, were genes. Isolating those genes and thereby providing the company something to show for itself in its earliest years, as well as giving it direction for the future, bestowed a status on the founding molecular biologists and their corresponding scientific discipline at Calgene that was not easily shaken.

Those of us selected to carry out the company's regulatory science were very much a part of the scientific culture at Calgene, whatever its underlying causes. And we knew before we picked up a single test tube that demonstrating the safety of the company's genetically engineered products, as opposed to discovering new genes and furthering scientific knowledge of gene expression or plant physiology, would occupy the bottom rung of Calgene's scientific hierarchy. Our scientific colleagues would not have much respect for our efforts.

Don Emlay, despite his Bachelor of Science degree (in entomology), was definitely not part of the scientific culture at Calgene. Having spent 13 years at Zoecon Corporation, part of that time in international marketing, he was a member of a completely different culture: the business culture. He enthusiastically described the science we were embarking on as the most important in the company. Don seemed genuinely excited about the challenge ahead, and by the end of his pep talk everyone in the room had to agree with him: regulatory science was the most important science Calgene scientists could be doing.

But it wasn't Don's call to meet the challenge that had convinced us. Rather, it was the stark realization that, without the regulatory approval we were seeking and the subsequent successful commercialization of the Flavr Savr tomato, the stability of the entire company was in jeopardy. What we were really undertaking was science for survival. We had to carry out regulatory science in the short term if we were to have any chance of getting back to science we found more interesting in the long term.

And, it turned out, we were talking extremely short term, at least for that first regulatory science effort. The first thing we learned about our initial regulatory science project was its due date. We were given just 3 months to design, carry out, and document all the necessary safety experiments to demonstrate that Calgene's selectable marker gene and its protein product were safe to use and consume in the company's genetically engineered products. Don planned to file the document with the FDA by the middle of November.

This method of deadline setting was in direct contrast to the way science had been done at Calgene up until that point. Until then, science at Calgene had been conducted more or less the way science was conducted at academic institutions. Even contract research, like that carried out for the Campbell Soup Company, was done "academically." Quarterly goals were set for each individual project and approved by Campbell representatives, themselves scientists. Good faith efforts were then made to attain those goals between quarters. But it was not at all unusual for the status of various goals to remain "in progress" at the end of the designated quarter and for the goal for the new quarter to be listed as "continue to . . . ." This was simply the reality of "life on the edge of technology." All the pitfalls inherent in the "discovery" phase of a scientific project could not be foreseen. Rather than trying to come up with a last-minute, slipshod answer, researchers would present evidence at the quarterly review meeting of an encountered pitfall and propose methods of getting around it. We often learned from the pitfalls themselves, and various new techniques and tricks of the trade came about as a result. The scientists we worked with at Campbell Soup had no problem with this scientific reality. Bill Reinert, the Campbell scientist in charge of scientific interaction with Calgene, once told me he was especially pleased with the science Calgene had done on Campbell's behalf, more pleased than with studies done by U.C. Davis scientists with whom Campbell Soup had also engaged in contractual arrangements.

In light of the way we had been doing science at Calgene, it was shocking to be told that we would file our document with the FDA, essentially submit our data for publication, no matter what, in a mere 3 months. If this assignment wasn't "edge of technology" science, complete with pitfalls and setbacks, what was? Not only had no one ever asked approval for a selectable marker gene in a whole food product before, but it would require expertise in technologies outside the experience of Calgene's team of researchers. Cell biologists would be doing microbiology. Plant biologists would need to become experts in human gastrointestinal bacteria. I tried to argue that the deadline was unrealistic, that it would undoubtedly have to be flexible. I looked to Bill Hiatt, the senior scientist in the room, for support. Bill agreed with Don and his deadline instead. I assumed at the time that Don had set the deadline and that it was just a taste of what it was going to be like to mix Calgene's business and science cultures.

I learned the even more disheartening truth years later when Bill revealed that it had been he, not Don, the regulatory bureaucrat, who had insisted on the 3-month deadline. "If I hadn't set a short deadline so that [everyone] could see the light at the end of the tunnel," he told me, "nobody would have wanted to participate in the project at all." Such was Bill's assessment of the powerful influence of Calgene's scientific culture.

In spite of the culture shock we were experiencing, the attitudes of the people brought together that hot August day were, for the most part, positive. The tomato scientists had been working together for at least 2 years, some 5 or more, and the group had gelled together nicely as a team. We knew we had an important, even if scientifically distasteful, job to do. So, with that deadline looming over us and with as much enthusiasm as we could muster, we broke open our copies of the ENVIRON document and got to work.

Choosing a Format

The first and most basic issue we addressed, after deadline setting, that is, was picking a format to use for data submission. At the time, the FDA did not have a specific process in place for dealing with genetically engineered whole foods.* It was up to us, therefore, to decide just how to submit to the agency whatever safety data we would produce.

The scientists took a logical approach. Most, myself included, believed the FDA's food additive petition (FAP) format was appropriate. And other scientists, like Rebecca Goldburg, with Environmental Defense (formerly, the Environmental Defense Fund), it turned out, felt the same way. Our selectable marker gene was expressed in every cell, including all the edible parts, of all our genetically engineered plants. Expression of the gene resulted in production of an enzyme responsible for rendering plant cells containing it resistant to several antibiotics. That enzyme is not normally present in tomatoes or any of Calgene's other potential products. Therefore, the enzyme, if not the gene responsible for its production, was a food additive.

Don Emlay could not have disagreed with us more. His arguments had a different kind of logic. He was worried

---

* Despite the agency's "Statement of Policy: Foods Derived from New Plant Varieties," published in 1992, it can be argued that it still doesn't.

about the required tests and procedures and associated costs that went into filing the standard FAP. Calgene couldn't afford that kind of money or that kind of time. He told us that "receiving a response from the agency for a food additive petition took at least 3 years!" Don preferred the advisory opinion option, and, I found out later, the FDA had actually advised him "to submit initially the information to be reviewed as a request for advisory opinion."[5] Advisory opinions are requested of the FDA in matters of general applicability. And since the selectable marker gene Calgene was using had broad applicability in the ag biotech industry—nearly every other company in the ag biotech business was using it—the advisory opinion route did appear to be an appropriate alternative.

Don also wanted to avoid using the *food additive* moniker and instead describe our selectable marker gene as a processing aid. He gave several reasons to support this definition. First, as was described in Chapter 2, the primary function of the selectable marker gene in plant genetic engineering was to increase the efficiency of identifying the relatively few plants that integrate new genes into their DNA. In this sense the selectable marker gene did serve as an aid in the process of genetic engineering. Second, Calgene was in a hurry. Obtaining regulatory approval for a processing aid was historically faster than for a food additive. And it was not legally necessary to publish safety data gathered for processing aids or include them on food labels. Therefore, describing the selectable marker gene as a processing aid kept our options open.

A brief debate ensued, during which we unanimously agreed that in light of the high level of public interest in the federal approval process for genetically engineered foods, we would make all of our supporting data public no matter what. We also decided to request an advisory opinion of the FDA for our processing aid. For lack of a better format, however, and in the spirit of compromise, we would submit our request in the FDA's food additive petition format.

I was comforted by our decision to work from a standard format, any standard format. Even a rough map would be better than nothing to help guide us through the uncharted regulatory waters. And as we launched into the first section, entitled "Identity of the Food Additive" (read, "Processing Aid"), I started to breath a little easier. I suspected that we had more, especially molecular, information on what our "food additive" was and where it came from than the FDA probably cared to know.

Hard-and-Fast Molecular Facts

The selectable marker gene Calgene used (and most of the ag biotech industry was still using at the turn of the century) is commonly referred to as a kanamycin resistance gene or neomycin resistance gene, abbreviated *kan*[r], *neo*[r], or *nptII*. The *kan*[r] gene encodes an enzyme called neomycin phosphotransferase II (NPTII), the specific form of which is called APH(3')II.[6] The complete DNA sequence of *kan*[r] and the amino acid (protein) sequence of APH(3')II have been known since 1982.[7]

APH(3')II catalyzes the energy-dependent phosphorylation of aminoglycoside-type antibiotics. Phosphorylation of these antibiotics interferes with their uptake into cells and their subsequent binding to ribosomes, the cellular structures responsible for protein synthesis. Therefore, cells that produce APH(3')II are resistant to the antibiotics that APH(3')II phosphorylates (kanamycin, neomycin, and geneticin, or G418, an antibiotic only used for in vitro experimentation) because those antibiotics can't enter and interfere with protein translation on the ribosomes of those cells. Other naturally occurring bacterial enzymes, such as acetyltransferases and nucleotidyltransferases, as well as other phosphotransferases[8] can also inactivate kanamycin and neomycin.

In addition to these intimate details describing our processing aid, we knew exactly where it came from. Calgene's kan[r] gene encoding APH(3')II had been isolated from a particular strain of *Escherichia coli* K12.[9] Further, it had been a component of the transposon Tn5, a so-called jumping gene.

While it was satisfying to be able to provide hard and fast facts about the *kan[r]* gene and APH(3')II for the FDA, it was worrisome as well. The bottom line was that our processing aid not only conferred resistance to antibiotics used therapeutically on humans but also had been part of a jumping gene isolated from a bacterial species the public knew and feared. Any PR related to these particular facts could not be good. But each and every prototype transgenic plant at Calgene—and there were thousands of them at the time—had that same *kan[r]* gene inserted into its DNA. We did not discuss these issues aloud. Rather, we seemed to silently agree that there was no looking back.

We used the next sections of the FAP, dealing with the use and intended technical effect of our processing aid, as a forum for presenting a short course on the procedures and techniques of plant genetic engineering. This explanation of the technology served several purposes. First, it was background information that put use of the *kan[r]* gene in perspective as a processing aid, as Don had suggested at the start of our first in-house regulatory science meeting. Second, since the entire advisory opinion package would eventually be made public, it could serve to educate U.S. citizens interested in learning about the technology. And third, it was intended to demonstrate the precision of the technology, especially as opposed to traditional breeding methods.

Detailed but Descriptive:
The "Precision" of Plant Genetic Engineering

As was described in Chapter 2, the techniques of plant transformation and regeneration can be used to isolate a single specific gene from, theoretically, any organism and insert that gene into the genome of, theoretically, any recipient plant. Plant breeders, on the other hand, are limited in their ability to insert genes into recipient plants by several factors. For one thing, because sexual crossing is their method of gene introduction, they can transfer genes only between plants that are sexually compatible. Introducing a tomato gene into a cotton plant, for example, would be impossible for a plant breeder but has been successfully accomplished by plant genetic engineers.

Traditional plant breeding also depends on genetic recombination for transfer of genes between parental chromosomes. The breaks in chromosomes required for these genetic exchanges generally occur randomly between as well as within genes. Because a plant breeder has no control over the location of these chromosomal breaks, the transfer of a gene of interest is inevitably accompanied by the transfer of additional genes and/or parts of genes. These hitchhiking genes are generally of unknown and sometimes deleterious function. A plant genetic engineer, on the other hand, knows each gene inserted into a recipient plant down to its individual nucleotide base pairs.

What we did not emphasize was that plant genetic engineering has, as opposed to the precision inherent in the theory, its own limitations. For example, plant breeders, and especially traditional geneticists, often know more or less the chromosomal location of the genes they want to transfer into their crops. And because those genes are bred into related plants through a recombination process that depends in part on like DNA finding and interacting with like DNA, they usually end up in a predictable, corresponding chromosomal location in the recipient crop plant. In contrast, plant genetic engineers at Calgene and throughout the rest of the ag biotech industry have no idea where their genes will end up in the DNA of a recipient plant. So not only did we have no control over where the Flavr Savr and *kan*[r] genes landed in a transformed tomato plant's DNA, but often multiple copies of the genes ended up at multiple sites in the recipient plant's chromosomes.

Insertion could even take place in the middle of a gene, disrupting and mutating it. Any phenotypic—that is, visible or physical—changes resulting from such insertional mutagenesis would be in addition to those expected as the result of expression of the inserted genes themselves and are therefore referred to as pleiotropic effects. But, because our request for an advisory opinion on the *kan*[r] gene did "not concern the safety of a specific commercial product, concerns relating to specific putative mutations that may be caused by the insertion of genes"[10] could not be addressed. It was at this point that trying to demonstrate the safety of hypothetical products started to get difficult.

A short appendix on pleiotropic effects that admitted that "pleiotropic effects can and do occur with regularity"[11] during the genetic engineering process was included in the original *kan*[r] advisory opinion document. But, the argument went, "insertional mutagenesis due to genetic engineering is similar to pleiotropic effects seen with conventional breeding."[12] And, since pleiotropy had not been a problem during centuries of classical breeding, we did not expect it to be a problem with genetic engineering either. We suggested that through "selective breeding and quality control" we could "assure a safe product for the consumer."[13] And, although some of us felt uneasy about devoting such a terse discussion to this important safety issue, within the confines of a generic document our hands were tied.[*]

The Limits of a Generic Document

The issue of the insertion of multiple genes at multiple DNA locations was also difficult to address generically. We didn't muddle our "genetic engineering is precise" argument by trying to deal with the issue there, however. Instead, we dealt with it as it became necessary, in following the FAP format, to provide factual answers to specific questions about our processing aid. We realized that, in order for the FDA to derive an opinion about and advise us on the safety of the kan^r gene and APH(3')II in foods, even a generic document had to provide some hard numbers. To make simple estimates for things like the human estimated daily intake of the food additive as well as to carry out probability analyses to assess any risk involved in their use, the FDA needed to

---

[*] In May 1992, the FDA concluded in its "Statement of Policy: Foods Derived from New Plant Varieties," that pleiotropic effects associated with plant genetic engineering were not a safety concern. But, unfortunately for Calgene, by then we'd freed our hands in order to deal specifically with the Flavr Savr tomato and gotten ourselves deeply mired in a pleiotropy-generated mess (see Chap. 5).

know how many kan^r genes and how much APH(3')II would be in our products.

Theoretically, we pointed out, only one correctly functioning kan^r gene was necessary to select transformed plant material.[14] Practically, however, we knew many of our transformed plants would contain more. But the higher the number of genes, or amount of APH(3')II, used for a risk assessment, for example, the greater any final risk would be. We had to establish upper limits. Those limits turned out to be one of the most controversial "policies" Calgene ever made.

In house, setting these limits was more of a problem for some programs than for others. It was not a big deal for the tomato program, for example. One of the first tests a transgenic tomato underwent at Calgene was a genetic determination of the number of places in the plant's DNA that genes had been integrated into. Because it was much easier to produce hybrid tomato varieties with plants that had genes inserted at only one chromosomal location, transformed plants with genes inserted at more than one place were discarded. Even though more than one gene was often inserted at the same genetic locus, this genetic screen greatly reduced the average overall number of genes present in a potential tomato product.

For Calgene's canola program, limiting the number of inserted genes was a bigger deal. Transgenic canola plants were selected for further development based on the extent to which their oil characteristics were altered compared to those of their nontransformed parent plants. Often, but not always, plants with the most desirably altered oils contained multiple copies of the inserted oil-altering, and the physically adjacent kan^r gene, integrated at multiple sites in the plant's DNA. Understandably, scientists in the canola program didn't want gene number limits imposed too strictly for fear that their best potential products, in terms of the genetically engineered trait, would have to be eliminated from development.

Likewise, Calgene's competitors in the ag biotech industry didn't like the idea of limits being set for *kan*[r] gene number and APH(3')II protein amount. They didn't want any decisions made at independent little upstart Calgene that might adversely affect their future products. But several of the largest of these companies had already had their chance to affect Calgene regulatory policy. Don Emlay had initially invited them to join our effort to gain the FDA's blessing for *kan*[r]. All had refused. Now they just had to hope we didn't mess things up.

But, controversy or no, Calgene's regulatory science team could not have done its job without establishing limits on *kan*[r] gene number and APH(3')II amount. A cursory survey of transgenic plants at Calgene was made. Somewhat arbitrary limits were then set. Any plant chosen for commercial development had to have ten or fewer copies of the *kan*[r] gene, and 0.1 percent or less of that plant's protein could consist of APH(3')II.[15]

With upper limits for the gene and its protein product established, Calgene scientists were nearly ready to hit the labs. We had the numbers we needed to establish, for example, human estimated daily intake for the kanamycin resistance gene and APH(3')II. But for what foods? Could or should we establish estimated daily intake for every possible food product Calgene might ever commercialize? It seemed impossible to be all-inclusive, especially given the generic nature of the document. And most of those putative products—soup, ketchup, canola oil, tomato paste, cottonseed oil, margarine—would be intensively processed. The chances of those processed foods containing much of our processing aid were minimal. Mechanically crushing seeds for oil liberates enzymes that degrade nucleic acids and proteins. The high temperatures used for processing oils and canning tomato products denature and degrade DNA and denature and inactivate enzymes. The low pH of processed tomato products, mandated by law to be 4.6 or less, would make an intrinsically hostile environment for *kan*[r] or any other DNA; DNA is depurinated and denatured at pH 4.5.[16] APH(3')II also would be inactive in processed tomatoes because its pH optimum is much higher than 4.6. Although we concluded that it would "be necessary to determine the actual levels of the gene and gene product in finished food products for human and animal consumption on a case-by-case basis"[17] (the FDA eventually found it unnecessary to do so, as described in Chap. 6), we eliminated processed or cooked *kan*[r] or APH(3')II food sources from further consideration based on these arguments.[18] The one food we were left with was fresh tomatoes.

The Safety Investigations

As the 20 or so of us sat in that crowded conference room reading over the draft of "Section E: Safety Investigations," the original nervousness I'd felt at the start of the meeting crept up on me again. Don, his experts from various panels, and staff at ENVIRON had defined three main areas of concern associated with the use of *kan*[r] and APH(3')II in plant genetic engineering. But because "risk assessments are always limited by the questions that one can think to ask,"[19] I knew we were back in virgin territory.

Don and his experts saw one major area of concern as the potential for APH(3')II in genetically engineered tomatoes to render ineffective the kanamycin a person might concurrently ingest to combat an infection. The other concerns had to do with the stability of the introduced gene. The problem was not so much that the gene might jump around within the DNA of the recipient organism and its heirs like the transposable elements originally observed by Barbara McClintock in corn. It was related instead to the remote possibility that the inserted gene might somehow migrate from the tomato or cotton or canola plant into other, sexually incompatible organisms via a process called horizontal transfer. Some bacterial species, for example, were known to be naturally transformed; they could, under very specific conditions, take foreign DNA up into their cells and incorporate it into their own genetic information. Could kan^r escape from a tomato during digestion in the human gastrointestinal tract and somehow transform bacteria residing in the human gut or the human epithelial cells lining the gut, making them resistant to kanamycin? And could such a gut transformation scenario occur in animals used for human food?

At first glance, tackling these issues seemed a gargantuan task, but Don explained that we would use a new conceptual framework to assess these potential risks: cascaded safety analysis. The overall risk would be broken down into smaller, well-defined steps that could be more accurately quantified and/or estimated. When multiplied together, the probabilities of these smaller steps occurring would represent the overall risk. It was a great idea and, I believe, the only way we could have approached the job we had before us.

But an approach is only as good as its implementation. And as the first meeting of the regulatory science team at Calgene was adjourned that August afternoon, it was time to get out and implement. We received our individual assignments as we left the conference room. Mine was to assess the risk inherent in the food-animal exposure scenario. I went back to my office in a state of mild shock.

As I skimmed through the information on the food-animal exposure scenario in the draft document, my shock turned more to dread. In addition to the risk of horizontally transferring *kan*^r to intestinal microflora of food animals and thereby decreasing the efficacy of veterinary antibiotic therapy, two other concerns were raised. Could animal intestinal microflora that were pathogenic to humans become antibiotic resistant and thereby serve as a source of decreased efficacy of human antibiotic therapy? And could the cells lining food-animal guts become antibiotic resistant? If so, ingestion of sausage casings, for example, could become another source for human exposure to *kan*^r. I spent the next 18 hours reading ENVIRON's white papers on "Food Animal Digestive System Physiology and Microflora," "Animal Feed Composition," and "Subtherapeutic Antibiotic Use in Animals" or dreaming about mutant bratwurst.

But the next morning, just as I was about to go bury myself in U.C. Davis' veterinary library, my assignment was changed. The food-animal scenario, it had been decided, would not undergo a full-fledged cascaded safety analysis. I was to work out the human exposure scenario for *kan*^r instead. Numerical estimates of the risk of gut microflora or epithelial cell transformation in humans would be used to indicate the extent of the analogous risks in food animals. The human scenario numbers might even overestimate the food-animal numbers, since many food animals have an extraruminal chamber where additional breakdown of nucleic acids occurs. And besides, sausage casings were extensively washed, which would remove most of the epithelial cells, transformed or not, and cooked. They therefore fell into the same category as processed tomato products and canola and cottonseed oils and would not be specifically experimentally addressed in our generic document.

I didn't question this decision. I was simply relieved. Being a plant scientist, my knowledge of the human gastrointestinal tract and its resident bacteria was rudimentary at best. But rudimentary knowledge is better than next to none, which was how much I knew about the gastrointestinal tracts or the intestinal microflora of food animals. I returned to my office with renewed enthusiasm.

The first thing I needed to know was the overlap between bacterial species that could undergo natural transformation and bacterial species that reside in the human gastrointestinal tract. Only one bacterial genus was present on both lists: *Streptococcus*. This was, I figured, a good news, bad news, situation. On one hand, the *Streptococcus* genus was well studied. The first demonstration of natural transformation, in fact, was with *Streptococcus* in 1928.[20] Therefore, I had more than 60 years of scientific data to use for assigning real numbers to the small, well-defined steps in my risk scenario. On the other hand, human diseases caused by *Streptococcus* bacteria were relatively commonplace. The perceived threat of *Streptococcus* specie's becoming antibiotic-resistant as the result of the use of plant genetic engineering would probably not be well-received by the public. Public relations aside, *Streptococcus* was what I had to work with.

Next, I needed to decide where in the human gastrointestinal system there were sufficient numbers of bacteria and adequate time for natural transformation to take place. Normally it takes 3 to 5 hours for stomach contents, or chyme, to pass from the pylorus, marking the border between the stomach and the small intestine, to the ileocecal valve, marking the border between the small intestine and the large intestine.[21] Chyme is sometimes blocked for another several hours in the lumen of the lower small intestine at the ileocecal valve until the person eats another meal,[22] and most of the resident bacteria found in the gut are just below the level of the ileocecal valve in the colon. Based on the hostility of the environments and lack of proximity to resident gut bacteria, I eliminated the stomach and upper small intestine and chose the lower small intestine, large intestine, and colon as the locations in the human gastrointestinal tract with greatest transformation potential.

I extracted from the scientific literature most of the numbers I needed to figure out the chances of an effective transformation event at that site, such as mean human consumption rate for fresh tomatoes, amount of DNA per tomato cell, volume of liquid in the lower small intestine and large intestine, and DNA recombination frequencies. But, though I scoured the published data for specific numbers on the survival rate of DNA in the human gastrointestinal tract, I found none. I knew it was generally assumed that, because pancreatic fluids are so full of enzymes that break down DNA, any DNA that had undergone human digestion would likely be nearly completely broken down into pieces no longer than one or several base pairs. I even found a reference that spelled that assumption out, although it gave no primary numbers or other references to back it up. The trouble, it seemed, was that this assumption was so generally believed that no one had actually done an experiment to demonstrate it. Obviously, I had to carry out those experiments myself. But, in the meantime, as I started writing my section of the document, I quoted from the general reference I'd found. It certainly couldn't hurt, I noted at another one of our regulatory science team meetings, to back up our in-house data with a public, non-Calgene–connected declaration of what the scientific community obviously held to be essentially self-evident, even if it did lack primary data. Bill Hiatt strongly objected to my use of any reference that lacked primary scientific data, however, and told me so in no uncertain terms in front of the whole group. More than unprofessional, he believed it would border on the unethical.

I'd been working with Bill for nearly two and a half years by then. Our work relationship had evolved from one in which I needed to prove myself, earn his respect, to one in which we were frank with each other to the brink of name-calling. He had pulled this unethical bit on me once before when I'd wanted to publish a paper without explaining a complicated, unexpected side observation. I argued that the observation deserved an entire paper of its own but to no avail. He refused to sign off on my manuscript. I went back to the lab and carried out the necessary additional experiments, but I was very unhappy about it. Six months later, my revised manuscript in hand, he finally, sheepishly backed down. "You're going to be really mad at me," he said. "You were right. The new experiments detract too much from the rest of the story. I think you should take them out." But how could I stay angry with a boss who could so readily admit he was wrong?

I was furious with him, however, over his remarks about that DNA degradation reference. As I was shooting basketball that night in an effort to diffuse my frustration with him, I tried to rationalize his actions. I was a particularly good shot when I was miffed at Bill, and on that occasion I couldn't miss. Maybe, especially since he chose to have the argument in a public forum, he'd wanted to make a general point about the scientific integrity he expected on the project. Or perhaps he was worried that I might decide I didn't need to do DNA degradation experiments after all. For all I knew, he might have just needed a good fight to relieve the tension that had been building up on the project as the self-imposed deadline closed in on us. Whatever his reasons, he apologized the next day, in a general way, and all was forgiven.

Fortunately for me, published recipes were available in the *United States Pharmacopeia*, the *National Formulary*, for simulated human gastric and intestinal fluids. Unfortunately for me, the smell of those solutions made me persona non grata in the lab during my experiments. Nonetheless, I made up these "juices," added long pieces of DNA to them, and, after various incubation periods and using sensitive radioactive detection methods, measured the length of what remained of the DNA. After 10 minutes in simulated gastric fluid and another 10 minutes in simulated intestinal fluid, conservative periods of time for the workings of the human gastrointestinal system, I could no longer detect DNA the length of the selectable marker gene or longer. Because a gene should be full length in order to be fully functional, this was a very reassuring but completely expected result. What's more, enzymes that break down DNA are also present in human saliva, and, although I had no formula for it, I conducted an experiment or two using the real thing. (A scientist has to do what a scientist has to do, and sometimes a scientist has to spit into a test tube in order to conduct a thorough experiment.) DNA was extensively degraded in that solution as well.

I was excited and relieved that I had the results I needed to complete my human exposure scenario. I'd read my x-ray films after regular work hours but decided to call Bill

Hiatt anyway to tell him the good news. I'd never called him at home before, and, as soon as he got on the line, I was sorry I had. He didn't sound happy to hear from me.

"The DNA's all broken down to half the size of the gene or less," I told him. "Even if a transformation event occurred, the gene wouldn't be functional."

"Great," he replied, "but that doesn't help us. We need to come up with a probability that the intact gene would be available for transformation. Reinterpret your data. See you tomorrow."

Bill was taking the ultraconservative approach: assume some genes, somehow, came through the digestion process unscathed. So, with the wind out of my sails, I went over my films again. I estimated, based partly on the limits of detection for my experiment and partly on pure conservatism, that 1/1000 of the DNA from a serving of Flavr Savr tomatoes could survive passage through the human gastrointestinal tract and be available for transformation of *Streptococcus* bacteria in the human gut. I came in to work the next morning with the first draft of the human exposure scenario completed. I never called Bill at home again, however.

A Surprise

One of the biggest surprises we had during our 3-month push to demonstrate the safety of the use of Calgene's antibiotic resistance gene came from experiments in which bacteria were artificially transformed with the *kan*^{r} gene. This gene had been specifically engineered to work in plants, and we assumed that, even if it somehow got into a bacterium, the chances of its actually working in the bug were essentially nil. Ray Sheehy and Dave Schaaf, a cell biologist from Calgene's cotton group, found otherwise.

The genetic code is universal. The information in a gene that codes for a protein can be translated, with varying efficiency, into that protein by any organism. That is why the coding region of Calgene's selectable marker gene, although isolated from a bacteria originally, could be read by a tomato or any other plant. The DNA signals responsible for regulating a gene's spatial and temporal function in one organism, however, are not always recognized by other organisms. Genetic engineers have to customize the protein-coding region of a donor gene by adding appropriate regulatory DNA sequences so that the coding sequences can be read properly, at the right time, and in the right cells by the recipient organism. Calgene's kan^r gene had been altered so that it would be properly expressed in plant cells. Because it had been so altered, we expected that the intact, engineered gene would not be expressed in bacterial cells.

Ray and Dave tested this hypothesis. Using standard laboratory techniques, they transformed *Escherichia coli* and *Agrobacterium tumefaciens* with kan^r DNA. Under these "unnatural" transformation conditions, the DNA was essentially forced into the recipient bacteria. But, could those recipients express the DNA forced into them? Ray and Dave spread the transformed bacteria onto media containing kanamycin and looked for colony growth.

The design of the experiment was not perfect. They were looking for a negative result, and in such an experiment one never knows whether one has looked hard enough. It would have been very reassuring, however, for those petri plates to have been completely devoid of bacterial growth when they were pulled out of the incubator the next morning. Instead, Ray and Dave found that a few bacteria, some from each species, had survived in spite of the kanamycin and lived through the experiment.

Several theories were put forward to explain what had happened. The few surviving bacteria in the experiment could have rearranged (recombined) their DNA in such a way as to juxtapose a DNA regulatory region, or promoter, of their own next to the coding region of the kanamycin resistance gene, for example. "Genetic engineering" of this kind by a bacterium would allow that individual to survive life on the antibiotic. But, to determine whether this theory was true or whether there was some other explanation for the "failure" of this experiment would have taken longer than Calgene's self-imposed deadline for submission of the advisory opinion document allowed. Additional experiments to elucidate the original finding were therefore never conducted. We simply assumed, for the purposes of our risk assessments, that, if the kan^r gene ever did get into a bacterium in the human gut or anywhere else, it would be properly expressed and would confer antibiotic resistance to the recipient bacterium.

Wrapping It Up in the Lab

We made many additional conservative assumptions for our risk assessments. For example, the number of *Streptococcus* bacteria present in a human gut was unknown and, undoubtedly, variable. We therefore assumed that all of the aerobic bacteria in the human gut were *Streptococcus* species capable of being transformed with the kan^r gene. (Anaerobic bacteria were considered already resistant to kanamycin and therefore irrelevant to the risk assessment.) We also assumed that the very specific conditions required for natural transformation in *Streptococcus,* involving high bacterial density and concentrations of a specific bacterial protein, would be met in the harsh environment of the human gastrointestinal system. Other factors related to the ease with which the *kan*[r] gene could be integrated into bacterial DNA and still be expressed were also assumed to occur with 100 percent efficiency even though, to the best of our knowledge, the likelihood of most of these events ever taking place at all was closer to zero.

Even making these and many other very conservative worst-case assumptions, we estimated that, at average levels of consumption (fiftieth percentile consumption rate[23]), for every 1000 humans that ate Flavr Savr tomatoes (or every 380 humans for the ninetieth percentile consumption rate[24]) one gut bacterium might become resistant to kanamycin. And out of a population of nonanaerobic (potentially sensitive to aminoglycoside antibiotics) human gut microflora estimated at 10^{12} one didn't seem too bad. Also, examples in the literature indicated that large numbers of bacteria isolated from humans were already resistant to these antibiotics. Seventy-five percent of the *S. faecalis* bacteria isolated from humans during a study conducted in 1982, for example, were kanamycin resistant.[25] In light of these numbers, our estimates of the number of human gut microorganisms that might become resistant to kanamycin due to eating genetically engineered Flavr Savr tomatoes seemed insignificant.

While I was piecing together the risk scenario for the *kan*[r] gene in humans, Bill Hiatt and Rick Sanders were doing the same for the gene's protein product, APH(3')II. With the goal of demonstrating how much APH(3')II would be available to inactivate a dose of kanamycin consumed coincidentally with a plateful of Flavr Savr tomatoes, they examined the protein's activity (i.e., its ability to inactivate kanamycin and neomycin) in the same simulated gastric and intestinal fluids that I used to estimate DNA degradation. They couldn't detect any. In fact, only 1/1000 of the protein itself, inactive or otherwise, remained after 20 minutes in gastric juice.

They noted that a lack of cofactors and an energy source, ATP, to drive the phosphorylation reaction likely contributed to their inability to detect enzyme activity, and that this same lack of cofactors and ATP in the human gut supported their conclusion that the risk of inactivating a dose of kanamycin was minute.

The Armchair Arguments

Several other arguments were used to bolster Bill and Rick's experimental evidence. For example, kanamycin is administered orally only as a prophylactic preoperative bowel preparation, during adjunctive treatment of hepatic coma, and for treatment of intestinal infections.[26] Since people undergoing bowel surgery, in hepatic comas, or suffering from intestinal infections seemed unlikely to be eating many fresh tomatoes, genetically engineered or not, the risk of compromising their treatment appeared remote.

Also, the human diet, especially a healthy one containing lots of fresh fruits and vegetables, is chock-full of antibiotic-resistant organisms. A fresh salad with lettuce, carrots, celery, cucumbers, and tomatoes is actually a major source of these organisms.[27] Drinking water is another significant source of resistant bugs.[28] With all the kanamycin-resistant bacteria we eat and the many resident human gastrointestinal tract microflora already resistant, at least some of those bugs must be producing APH(3')II. And, if we're eating lots of APH(3')II already, maybe it shouldn't be considered novel at all, not for its possible inactivation of a dose of kanamycin and perhaps not even as a putative allergen.

Genetically engineered cells containing intact, transcriptionally active *kan*^r had also been recently injected into human cancer patients and children with severe combined immune deficiency. These experimental therapies, which also utilized kan^r as a marker gene, were just underway as we put together the kan^r advisory opinion document. Early indications were that no adverse side effects related to kan^r or to APH(3')II could be clinically observed. Calgene believed that these gene therapy studies provided "perhaps the strongest arguments for the safety of the kan^r gene in humans."[29] If active copies of kan^r could be injected into humans and cause no adverse effects, then surely the minute exposures anticipated from food crops, especially combined with the fact that both kan^r and APH(3')II would be almost entirely degraded by the digestion process, should result in no apparent ill effect or toxicity from eating Flavr Savr tomatoes.

However, Environmental Defense found the preliminary results from the gene therapy trials a particularly weak argument for kan^r safety. Rebecca Goldburg and her staff noted during the public comment period on Calgene's document that the desperately ill condition of the five patients in the cancer trial "may have masked any responses to the marker."[30] They went on to say that, while taking unstudied risks may be acceptable for sick individuals, it is "not appropriate for the members of the general population who do not need the marker in their food."[31]

Pulling the Final Document Together

The couple of weeks before the document was filed were especially hectic. Some of us gathered, digested (so to speak), wrote up, and sent our contributions to ENVIRON staffers, who were assembling the entire package, near the last minute. I was especially guilty. I finished my appendix on the frequency of gastrointestinal bacteria transformation late at night the day it was due. Because we'd had formatting problems due to software incompatibilities between ENVIRON and Calgene, Matt Kramer helped me fax the appendix to ENVIRON.

Bill Hiatt could barely believe I'd made the deadline when I told him the next morning. He even shook my hand, a gesture he usually saved for occasions like promotions or birthdays. Later that day, however, he got a call from ENVIRON. Did Calgene really expect them to type such a nasty technical appendix into the document under the current deadline? Don saved the day by hiring temporary secretarial help. The final touches were put on the document over the Thanksgiving weekend.

With all the rush, there was little time to review the document before sending it off to the FDA. The highest-level scientists at Calgene who weren't otherwise involved in the project were given copies for their approval, and a few of us on the regulatory science team were asked to review sections we hadn't contributed to ourselves. But there was barely enough time to do a decent editing job on the entire package, much less suggest substantial changes. And I found the environmental assessment (EA) volume of the kan^r advisory opinion especially in need of substantiating.

In an especially ambitious move, Calgene also filed an EA as part of its request for an advisory opinion on the safety and use of the kan^r gene in the production of genetically engineered plants. With foresight, back in March 1989, the idea had been to first file for and receive a formal exemption from the permitting requirements for field tests of genetically engineered Flavr Savr tomatoes from the U.S. Department of Agriculture (USDA). The same environmental assessment information, accompanied by the USDA's official ruling that it no longer considered the tomato a regulated article, would then be filed with the FDA, which had expressed interest in having the USDA rule on the broad-scale environmental issues related to plant genetic engineering. A USDA exemption would have required specific information on specific crops, however, and since Calgene's regulatory strategy had in the meantime become "generic," the original plan had been abandoned. Why assessing the effects of multiple hypothetical products on numerous generic U.S. environments (e.g., 21 U.S. states were identified as the generic environments in which tomatoes would be grown) was considered the superior course of action remains unclear. Nonetheless, the company made the attempt.

Many aspects of the EA were repetitive of the FAP part of Calgene's request for an advisory opinion. Details of the kan^r gene DNA sequence, for example, and arguments related to the amount of APH(3')II already in the environment and the safety of human gene therapy with kan^r were given again. But three main types of environmentally relevant information distinguished the EA from the FAP. The first was a cascaded safety analysis of the possibility that soil bacteria might become transformed by the kan^r gene DNA present in agricultural waste. The second was a discussion of the cross-pollination capabilities of tomato, cotton, and canola. And, finally, copies of official government (USDA) documents stating that no significant impact on the environment had been observed during various field trials of genetically engineered crops were included in the EA as appendixes.

The cascaded safety analysis was performed by Catherine M. Houck, Ph.D. Cathy, a chemist turned molecular biologist, had been hired at Calgene not long after Bill Hiatt was brought on board. She had been involved with tomato research since the initial contract with Campbell Soup in

1985 but had left the tomato program at the end of the previous fiscal year, when Calgene had acquired Plant Genetics, in order to head up a genetic engineering effort in alfalfa. The latest company refocusing had eliminated the alfalfa program, however, and set her adrift. Bill Hiatt, kicked further upstairs into management as a result of the same refocusing, considered putting Cathy in charge of the day-to-day aspects of running the tomato program. But his timing on the idea was bad: he came up with it a few days after he'd already given the added tomato responsibilities to me. At that point, with a guilty look on his face, he asked me what I thought about letting Cathy take charge of tomato. "Not much," I said, and he left my office without further discussion. Cathy eventually found her niche heading the development end of the cotton R&D program. (David Stalker led the research end.) For a large part of the next year, however, Cathy's new job, and mine as well, meant "regulatory science."

As a result of her analysis, Cathy estimated that anywhere from 1 or 2 to 900,000 (in an unlikely worst-case scenario) naturally transformable bacteria per hectare per year of agricultural land planted in genetically engineered crops could become kanamycin resistant by incorporating the Calgene kan^r gene present in decaying plant material in the field into their DNA.[32] She then compared this estimate to the number of kanamycin-resistant organisms assumed to be already present in agricultural soils, some 7.2×10^{12}.[33] She concluded that any newly resistant bacteria would constitute, at worst, not more than 1/10,000,000 of the existing resistant population[34] and would therefore be insignificant.

But more than the theoretical worry of soil bacteria becoming transformed with the kan^r gene was the all too real possibility of plant-to-plant transfer of the gene via cross-pollination. This gene flow issue is related to a potential for "weediness"[35] and was (and still is at this writing) of great concern to environmentalists. Of course, if there is no cross-pollination, that is, if the plants in question are only self-fertilizing, there would be no gene flow. To a greater or lesser extent, each of Calgene's core crops was described in the EA as self-pollinating: tomatoes "strictly,"[36] cotton "normally,"[37] and canola "predominantly."[38] The published numbers cited in the EA, however, told a different story. Significant levels of outcrossing had been observed for each species: 3 to 30 percent for tomato,[39] 0 to 50 percent for cotton,[40] and 5 to 95 percent for canola.[41] But even this level of outcrossing need not be troublesome as long as there are no sexually compatible relatives for the crop plants to outcross with. In the United States, this is essentially the case for tomato and cotton (with the notable exception of non–genetically engineered plants of the same species). Canola, on the other hand, crosses not only with other crop plants, such as rutabaga, fodder rape, turnip greens, turnips, and Chinese cabbage, but also with field mustard, a weed of considerable concern in the U.S. Northwest.[42] What's more, canola itself has "weedy" characteristics.[43]

In defense of the environmental safety of growing an estimated 1.0 million acres per year of genetically engineered canola in the United States, Calgene promised that the outcrossing and associated weed-related problems of canola would be "managed with crop rotation and good agricultural practices."[44] Unfortunately, however, specific examples of the good agricultural practices required for canola management, even on the small scale of a field trial, were not given in the EA. None of the field trials reported in the document had been of canola plants.

The other defense given for the environmental safety of canola transformed with the *kan*[r] gene also applied to tomato and cotton. As long as there was no strong selective pressure (e.g., available antibiotics) for developing and maintaining antibiotic resistance in an environment, plants producing APH(3')II should not have any advantage over other organisms in that environment.[45] In fact, published evidence suggested that the effort required to produce a protein, such as APH(3')II, that confers antibiotic resistance decreases an organism's fitness in a natural environment.[46] Without the advantage of increased plant fitness, Calgene argued that there was minimal need for concern over kan^r-containing plants in the environment.

I voiced concerns over the rushed review process and the general quality, especially of the EA, to Bill Hiatt when it came time to list the document's preparers. I told him I didn't want my name on it. I must have surprised him because he didn't say anything for a moment, which was very unlike him. When he did speak, it was only of my "significant" contribution. He insisted that I be listed as a preparer on the FAP volume. I backed down without much of a fight, but with the hope that next time we would do things differently, when he assured me that my name would not be on the EA volume. (And, in large part because of Bill, things were done much differently the next time out; see Chap. 4.)

As it turned out, I need not have worried about the completeness of the kan^r advisory opinion document. Like the editor of a scientific journal deeming a manuscript incomplete, the FDA, as a result of its own internal review and in response to comments it received from the public, was to ask multiple times for additional experimental data to support Calgene's conclusion that kan^r was safe to use in producing genetically engineered plants. We were not nearly finished with the kan^r gene story when we filed the advisory opinion document with the FDA on November 26, 1990.

The Celebration

Calgene's big press release that November day was not only picked up by the *Wall Street Journal* and the *New York Times,* among other newspapers, but also elicited an immediate response from Environmental Defense and the National Wildlife Federation. These environmental groups, generally regarded, at least at Calgene, as anti-biotech, actually praised Calgene for voluntarily coming forward and asking for government approval. They also chided the FDA for not having a specific program ready and waiting to deal with genetically engineered food products. "Are we going to depend on companies voluntarily coming forth?" asked Dr. Margaret Mellon, director of the National Wildlife Federation's Biotechnology Policy Center. "That's not enough for us. . . . As long as there is no legal obligation to do so, they are going to avoid the process of a lengthy government review."[47] Calgene officials scored the situation as an important PR coup.

The filing of the *kan*[r] FDA document and the initial public response it received caused spirits to run especially high at Calgene that holiday season. Bill Hiatt was so relieved that the document was away and grateful for the pressure-packed effort that his team of scientists had made that he threw a party. Since Bill usually avoided Calgene social activities (except those for which it was rumored Roger Salquist took mental note of staff attendance), a party at Bill's house was quite an occasion.

Bill had, however, his own idea of how to throw a party and, after a brief toast of thanks to his staff, spent the rest of the time in front of the tube watching the Big Game between Cal and Stanford. I shared his general party etiquette, and I had gone to Cal. I watched with him. He gave me a hard time for leaving early, before the game was over, but I had another engagement and was confident that Cal had the game in the bag. Bill reveled in giving me grief the next day over Stanford's recovery of an onside kick to set up and make a 39-yard field goal in the final seconds to win the game.[48] While the Cal loss was hard to take (Roger Salquist also flashed the "F*** Cal" button he wore hidden in his lapel at me), the interaction with Bill gave me hope that I had not only earned his respect for my science but also had a chance at admission into his "good old boy" club. Hope sprang eternal.

Even rarer at Calgene than parties at Bill Hiatt's house were bonuses for the science staff. Roger presented them to those of us who worked on the *kan*[r] document at an all-employee gathering a few days before Christmas 1990. The text of the accompanying letter read

> The filing of our *kan*[r] petition with the FDA this month was truly a landmark event for Calgene and for our industry in general. Your selfless, creative, and dedicated participation on the Calgene team responsible for preparing this document is most appreciated by the company and its shareholders. The dynamic process by which this submission was prepared and your enthusiastic participation in the intellectual give and take of defining this is something to be truly proud of.
>
> I am pleased to present you with this special Calgene award of $1000 in partial recognition of your personal efforts in the *kan*[r] filing.
>
> Congratulations,
> Roger Salquist

I thanked Roger after the ceremony. He said he hoped that I could get a little something extra for Christmas with the money. "More than the money," I responded, "I appreciate the public recognition." He seemed slightly ruffled, perhaps surprised, by my remark. I remember thinking that the business and the science cultures at Calgene were still worlds apart.

We also received kan^r team shirts at that preholiday celebration gathering (an extra supply of which Don Emlay kept on hand for people like himself who had a tendency to turn the white shirts pink in the wash). But my favorite token of appreciation for participation on the Calgene kan^r regulatory science team was, surprisingly, one taken from the business culture. Don gave us miniaturized versions of the submitted document embedded in Plexiglas™ blocks. I'd seen similar blocks, containing documents from various business deals, public offerings, and the like, on the desks of some of Calgene's business staff. While that block was an obvious physical sign of the convergence of science and business at Calgene, I liked it anyway. It still occupies a prominent place in my home.

CHAPTER FOUR

The Long,
Hard Regulatory Labor

The Flavr Savr Tomato Advisory Opinion

We didn't bask in the glory of filing that first request with the FDA for long. When I got to work after the New Year's holiday in 1991, a stack of papers 6 inches high was waiting on my desk. Calgene was wasting no time. The generic safety assessment of our selectable marker gene, *kan*[r], was off to the FDA. It was now time to specifically assess the safety of the first product in Calgene's genetic engineering pipeline: the Flavr Savr tomato. Obtaining approval from the FDA to sell it for general consumption was the company's top priority.

The New Team

From the outset, assembling the Flavr Savr tomato advisory opinion document was a considerably different experience than putting together its *kan*[r] predecessor had been. For one thing, the makeup of the regulatory science team personnel had changed significantly. Since the project was now tomato specific, the cotton program scientists, with the exception of Cathy Houck, had been released from Calgene regulatory science service and allowed to return to their former jobs. Likewise, ENVIRON International Corporation was excused from active involvement in the preparation of the Flavr Savr tomato advisory opinion document. Instead, and appropriate to the task at hand, the Flavr Savr tomato regulatory team consisted of Calgene's tomato program scientists and its regulatory affairs officials, Don Emlay and Keith Redenbaugh. In addition, we would utilize the knowledge and talents of our colleagues at Campbell Soup's Institute for Research and Technology. Calgene's somewhat disparate kan^r regulatory science team had been transformed into a lean, tomato-specific research machine.

The person most responsible for the team's transformation and its undeniable leader this time out was Bill Hiatt. Bill's frustration over the bicoastal logistical problems associated with ENVIRON was only one reason why he wrested control of the project from Don Emlay, who was, after all, Calgene's director of regulatory affairs. The other was related to the power struggle that had been brewing between Bill and Don.

The relationship between the two men had never been good. In many ways they were like night and day. Bill was a "man's man." When he wasn't talking animatedly about work, he was talking animatedly about sports or hunting. And yet he was also a family man, married to his high school sweetheart. He even passed cigars out at work to celebrate a new baby in his household (although not as a consequence of his youngest child's birth, but at the earlier news that the baby would be a boy). Don, on the other hand, was single at the time, although he did talk about his "kiddos" from various previous marriages, and soft-spoken. I found him on several occasions quietly eating instant oatmeal in Calgene's executive kitchen. The one thing the two men did seem to have in common, an admiration for classic cars, also, it seemed to me, appeared to define their differences. Bill drove a 1965 dark fatigue-green Mustang, Don a 1956 Chevrolet. (Don's admiration tended toward the obsessive, however; to vie for his attention, his kiddos occasionally resorted to throwing dirt clods at his car.)

Through the tense 3 months of putting the *kan*[r] document together, what little regard the two men had for each other had only deteriorated. The final straw for Bill may have been over whom deserved credit for the cascaded safety analysis approach used to assess *kan*[r]. At one of our last *kan*[r] regulatory science team meetings, Bill called the approach the central theme of the *kan*[r] advisory opinion document and proclaimed that it had been his idea originally, which was somewhat unlike him. Don laughed off Bill's proclamation, giving the credit instead to ENVIRON Corporation. Bill seethed instead of defending himself, and I can only assume that Don was right in the matter since the document, as filed with the FDA, states that the framework was "developed by ENVIRON."[1] After that confrontation, Bill and Don seemed to avoid each other as much as possible. More and more, Keith Redenbaugh was serving as the "go-between" that kept communication between the two men going.

Whatever Bill's reasons for taking the reins in the regulatory arena for his Flavr Savr tomato, those of us on his staff were all for it. Calgene tomato team members had great respect for and loyalty to Bill. We believed that, with Hiatt at the helm, the process this time around would be even more science-driven. And one early sign of that was Bill's deadline for the document. He was giving us until August, more than twice as long as we'd had to prepare the *kan*[r] document. Even devil's advocates like me believed an August deadline would give us ample time to do the job right. And the prospect of working with the scientists at Campbell Soup again made for a certain air of old home week, a return to the "good old days," around the project. We believed the tomato team was back on familiar ground with familiar leadership. We anticipated a much smoother ride.

The Game Plan

Although Calgene had no indication from the FDA that the information contained in its first request for an advisory opinion was adequate for the agency to come to a conclusion in the matter (the FDA had, after all, only received the nearly 900-page, plus references, *kan*^{r} documents a few weeks earlier), we nevertheless set out to produce the second document with a high level of confidence. We worked from the assumption that our quest for FDA approval of *kan*^{r} would be successful and that therefore we needn't deal with the selectable marker gene issue again. That assumption left us with a very innocuous genetically engineered product to seek approval for. If we ignored the complication of *kan*^{r} and its protein product, the genetic modification to the Flavr Savr tomato was essentially just an extra copy of a tomato gene reinserted into the plant upside down and backward. This upside-down and backward gene produced no protein product, which made the question of "processing aid" versus "food additive" irrelevant. Without *kan*^{r}, extra nucleic acid was the only thing added to the Flavr Savr tomato that set it apart from other tomatoes. And, since people eat nucleic acids like DNA every day in animal and plant food products, the Flavr Savr gene itself, we believed, should be "generally regarded as safe" (GRAS). Consequently, determining the format in which to submit our data to the FDA was easy. We would request another advisory opinion of the agency but use its GRAS affirmation format this time around.* In his letter to the FDA accompanying the *kan*[r] request, Don Emlay described that document as "in the food additive petition or GRAS affirmation format." This time out we would not seem so fickle.

Since we believed the Flavr Savr gene was GRAS, our mission became to demonstrate that Flavr Savr tomatoes were food just like any other tomato and should therefore be subject to the same regulation, and only such regulation, as other tomato varieties produced using more traditional methods. Of course, the simile could be taken too far. And Keith Redenbaugh, in a particularly zealous original draft of the first section of the Flavr Savr advisory opinion document, made that point early on.

Keith (and Don) took the idea that the Flavr Savr tomato was essentially the same as any other tomato quite seriously in the first draft of the "identity" section of the tomato document. That is, for Section A.1, "Common or Usual Name" (of the subject of the advisory opinion), he listed simply "tomato." For Section A.2, "Formal Name," he listed the scientific name of tomato, *Lycopersicon esculentum,* and so on and so forth. No mention of the fact that this tomato had been genetically engineered was made at all in the opening pages of the first draft. I objected. Wasn't this taking things a little too far? Couldn't the FDA take one look at this first section of the document and decide that the issue of whether

---

* Except for their titles, the food additive and GRAS affirmation formats are identical. Both require information in five general areas: (1) identity, (2) proposed use, (3) intended technical effect, (4) method of analysis of the substance, and (5) full reports of all safety investigations regarding the substance. They differ only in how the information presented is interpreted by the FDA as fitting the specific definitions established for each specific food category.

"tomato," described in no other explanatory terms, was in fact a food was not worth their consideration?

Don and Keith backed down after a short debate. There were, in fact, discernable differences between Flavr Savr tomatoes and traditionally bred tomatoes, differences that, after all, made our tomato better and were the basis of the business we would build, right? The trick, everyone agreed, was in how one presented those differences. So we changed the common or usual name from "tomato" to "Flavr Savr tomato" and its formal name to "*Lycopersicon esculentum* Mill, transformed with pCGN1436 containing an antisense polygalacturonase gene,"[2] thereby admitting that, yes, our food was different, but it was still food. And instead of claiming that our tomato was just like other tomatoes, we set out to compare and contrast the Flavr Savr tomato with tomatoes that had not been genetically engineered.

What few differences did exist between Flavr Savr and other tomatoes, the argument went, were all the result of specific intentional changes. Our objective became to document those intentional changes and, as far as possible, the lack of any unexpected surprises in our tomato and thereby demonstrate that the Flavr Savr tomato, although somewhat different from traditionally bred tomato varieties, was not significantly so. This focus on characterizing the "intended" versus "unintended" modifications made to genetically engineered foods remains the core of the FDA's general assessment approach to these new products as of this writing.[3]

The Intended Technical Effects of the Flavr Savr Gene

Matt Kramer, who had been in charge of most of the greenhouse and all of the field studies defining just what the Flavr

Savr gene had done to make Flavr Savr tomatoes different from their traditionally bred counterparts, put together Section C, "Intended Technical Effect of Flavr Savr Tomatoes." Matt described reduction in the amount of active polygalacturonase (PG) enzyme in ripening transgenic fruit as the "direct" intended effect of expression of the Flavr Savr gene.[4]

This direct effect had by then been demonstrated beyond a shadow of a doubt and in hundreds, if not thousands, of tomato plants. The relatively quantitative method used to demonstrate reduction in PG enzyme level was also straightforward to carry out and visually convincing. Literally anyone could conduct the test, called a PG assay, and select plants with reduced levels of PG from those with the usual high levels, and Calgene often employed undergraduate college interns to do so. Reduction in PG enzyme as the direct intended effect of the Flavr Savr gene (or similar antisense PG genes) had also already been published in reputable, peer-reviewed scientific journals, and not just by Calgene scientists. Matt cited plenty of previously published papers as documentation for the FDA of the direct intended effect of the Flavr Savr gene.

Documenting how that direct intended effect led to "indirect" intended effects on tomato fruit, such as longer field holding, increased serum viscosity, decreased susceptibility to various postharvest diseases, and, of course, improved firmness was dicier. It was the spring of 1991. The scientific controversy that surrounded the PG enzyme and its relationship to fruit softening (see Chap. 2) was still raging. How could Matt convince FDA scientists of these "indirect" intended effects when Calgene had so far failed to convince its scientific rivals of them?

Luckily, Matt had his pièce de résistance waiting in the

wings. The manuscript containing Calgene's support for its contention that reduced levels of PG enzyme could make ripening tomatoes firmer was written and under review for publication in *Postharvest Biology and Technology*. The timing of that paper could not have been better. It contained not only statistically significant data supporting the idea that Flavr Savr tomatoes soften more slowly than do nonengineered control fruit but also additional data demonstrating other pectin-related "indirect" yet "intended" effects of Flavr Savr gene expression in tomatoes as well. The manuscript, as submitted to *Postharvest Biology and Technology,* was included in its entirety in the Flavr Savr tomato advisory opinion document as an appendix to Section C.[5]

While the scientific peer-review process was only just under way for Matt's *Postharvest Biology and Technology* paper by the time the Flavr Savr tomato document was submitted to the FDA in August 1991, chances were good that the manuscript would be accepted for publication soon thereafter. (And it was accepted October 31 of that year.) Because the FDA preferred to have peer-reviewed scientific literature for its review process, Matt and the rest of us on the tomato regulatory science team cited Calgene's published results, as well as pertinent published results from other labs, as often as possible. Most of Calgene's publications on PG and tomatoes, in fact, were included as appendixes to the document. Having peer-reviewed published papers to work from made Matt's regulatory assignment of documenting the intended technical effects of Flavr Savr tomatoes relatively straightforward . . . except for one considerable glitch. All of the research carried out for every one of the publications cited in that document had been conducted, not with the Flavr Savr gene construction itself, the one that would be included in commercial Flavr Savr tomatoes, but with a first-generation prototype gene construction instead.

The molecular biology of the prototype gene construction was a mess. (For the purpose of this description, the term *gene construction* is synonymous with T-DNA.) It contained genes and parts of genes other than the antisensed version of the tomato PG gene and *kan^r*. It not only carried genes conferring resistance to the antibiotics chloramphenicol and gentamycin but also expressed those antibiotic resistance genes by virtue of bacterial gene promoters. And because it was such a mess, Roger had previously decided in a taxi with Bill Hiatt and Don Emlay on the way to a meeting with FDA officials, that it would not be commercialized. Calgene feared that using a gene that conferred resistance to an antibiotic such as chloramphenicol, used relatively frequently for human therapeutic treatment, in genetically engineered crop plants would never fly with the FDA.

As Calgene struggled with these issues in house, Don Emlay also tried to convince Ciba Geigy (now Novartis) not to commercialize an ampicillin-resistant variety of corn they had produced, but to no avail. The FDA has since allowed commercialization of that corn variety in the United States. Ampicillin-resistant corn has encountered resistance in the European community, however.[6]

Ray Sheehy had had the cleaned-up, "commercial" version of the Flavr Savr gene construction ready the previous year, but there was just not enough turnaround time to legitimize it through the peer-review publication process prior to our self-inflicted deadline for submission of the tomato data package to the FDA. Consequently, Matt had to settle for demonstrating the intended technical effects of the Flavr Savr tomato primarily with fruit containing the prototype gene construction. That didn't seem like such a bad compromise at the time because the "antisensed" version of the tomato PG gene in the prototype was identical to the Flavr Savr gene in the commercial version. (We were very careful to distinguish the prototype from the commercial version throughout the tomato advisory opinion document. Today, it is highly unlikely that the FDA would accept any supporting data produced using a gene construction other than the one destined for commercial use.[7])

When it came to assessing the safety of commercializing Flavr Savr tomatoes, however, the prototype gene construction was just not going to cut the mustard. Therefore, with the FDA as well as product development in mind, Hiatt's tomato group had been in high gear since January 1990 producing bona fide Flavr Savr tomatoes with Ray Sheehy's new, improved Flavr Savr gene construction. Bill wanted at least ten potential Flavr Savr tomato products to work with, and he wanted them characterized more completely than any transformed tomato plant—no, more completely than any tomato plant, period—had ever been characterized before. With luck, one of those ten would end up being our actual first product. Without luck, they would still showcase the precision of agricultural biotechnology and serve as examples of what our eventual product would be like.

By January 1991, 21,250 individual transformation "experiments" had been carried out. In 960 of them, regenerated and transformed plants had been identified by virtue of their having produced roots on media containing kanamycin.[8] The frequency of transformation varied from 20 percent for one tomato variety to 1 percent for a particularly recalcitrant cultivar.[9] Less than 20 percent of those regenerated, transformed plants had the 95 percent or greater reductions in PG enzyme level that were considered commercially viable. Of the 167 transformed plants remaining, only 130 produced enough seed to allow a genetic determination of the number of places in each plant's DNA that the Flavr Savr gene construction had been inserted into.[10] (This determination was carried out using a straightforward test in which seeds, the products of self-fertilization in tomato, were placed on growth media containing kanamycin. As per classic Mendelian genetics, which is based on segregation and independent assortment of chromosome pairs during meiosis, the ratio of seeds that germinated versus those that did not was an indication of the number of places at which insertion had taken place in the parent plant. A 3:1 ratio indicated insertion at a single locus, 15:1 at two loci, 63:1 at three loci, etc.) Forty-four plants carried the Flavr Savr gene construction at only one genetic locus, as determined using this test.[11] After more than a year, these 44 plants were all that we had to work with. We now had approximately 6 months to select and fully characterize Bill's ten shining examples from among them.

The Rate-Limiting Step

The task of selecting the ten Flavr Savr tomato plants for further characterization was assigned to me. Our selection criteria at this point were exclusively molecular. I needed to know exactly how many Flavr Savr and *kan*[r] genes had been integrated into each plant and verify that the insertion process had occurred as expected, that is, that the structure of the transferred DNA post-insertion was identical to its structure pre-insertion. The remaining safety analyses would be conducted on the ten plants I could most completely and easily explain molecularly.

While I was confident that I could generate the required data on gene structure and number and analyze and write it up by the August deadline for submitting our request to the FDA, I also realized that this analysis was the rate-limiting step for the entire project. The levels of APH(3')II, vitamins, putative toxins, etcetera, in Flavr Savr tomatoes could not be measured before my experiments were complete. Acute oral toxicity tests could not be conducted until the plants from which to harvest the Flavr Savr tomatoes that would be fed to rats had been selected. I needed help, I argued, to speed up my part of the project so that these other critical analyses would not be put on hold until the last moment. And although Rick Sanders formally (at least as formally as these things got at Calgene) reported to Matt Kramer at the time, Bill agreed that Rick should work with me.

Being responsible for the rate-limiting step created a certain amount of pressure, much of which we put on ourselves. But Bill Hiatt was by far our greatest source of pressure. I told him repeatedly that, even with Rick's help, the analysis would take several months, but all he ever wanted to know from me, at least several times a week, was whether we were done yet. I attributed his constant concern to two facts. First, he was not familiar with the techniques and concepts of DNA analysis that Rick and I were using. His expertise was protein, not DNA. Second, he just wouldn't sit still long enough to let me explain the experimental design. Out of frustration I prepared an hour-long presentation for one of our weekly staff meetings in which my plan was to educate Bill, as well as other members of the tomato team, about the experiments Rick and I were doing. I figured that if Bill understood all that was involved in the analysis, he'd give us the appropriate breathing room. I'd barely started my introduction to the presentation, however, when he interrupted.

"What's the bottom line?"

I gave him a 1-minute status report and went back to my introductory remarks.

"I don't want details," he said and moved on to other business. My educational seminar had been completely usurped.

As the August deadline for filing the document approached, things got even worse. I had long before planned and received permission from Bill to take a 2-week vacation that July. The "example" varieties had been selected by then, and myriad other analyses on those varieties were in full swing. Despite my attempts to reassure him that my sections of the document were written and all the figures and references ready to go, Bill was not easily consoled.

"I know I've already OK'd your vacation, but can't you postpone your trip?" he asked.

"This trip has been planned around a total eclipse of the sun and I believe that another one won't occur in the United States until the year 2017," I replied.

Slumping in his chair, he stared at me for a minute or so. I knew that he wouldn't be feeling so unsure about the whole situation if he'd just taken a few minutes, at some point during the previous 5 months, to walk through the design for the DNA analyses with me. Consequently, I wasn't about to make it easy for him. If he really didn't want me to go, he'd have to say so. I was not volunteering to forgo my vacation. Besides, I was returning a few weeks before the filing deadline. There would be plenty of time for any additional input I could contribute then. Finally, he wandered out of the room without another word.

When I got back from my vacation there was a note from Bill scribbled on a piece of scratch paper on my desk. It read:

Belinda,

Welcome back.
Great job. You and Rick have done the impossible.

Bill

During the 2 weeks I was away, he'd obviously read my report carefully and come to appreciate the power of the Southern blot technique Rick and I had used (see below). I carried that piece of paper around with me in my briefcase for several years. Bill didn't often hand out compliments like that one.

Selection by Southern

Rick and I utilized several variations of the Southern (named after the scientist who invented it[12]) blot–type experiment to complete the mission Bill deemed impossible. In a typical Southern blot experiment, DNA is isolated from individual plants, cut into various lengths with an enzyme or enzymes that cleave DNA only at specific sites, and pulled through a gelatin-like matrix based on its net electric charge in the process of gel electrophoresis. Smaller DNA fragments migrate farther into the gel because they travel through the gel matrix more easily and therefore faster than do larger DNA pieces, which stay nearer their point of origin. Thus separated by size, the DNA fragments are transferred to a filter paper membrane, again using an electric current or by simple capillary (wicking) action. This transfer of the DNA fragments is most often accomplished in the laboratory by placing the filter paper membrane directly on the gel and then placing a large stack of paper towels on top of the membrane. The liquid in the gel, and usually additional liquid provided beneath the gel, is drawn up through the membrane and into the absorbent paper towels in a matter of hours. The DNA in the gel goes along for the ride as far as the membrane "blot," where it is deposited in the same spatial array as it had been in the gel originally. Individual fragments of DNA can then be detected and their sizes determined by probing the blot with a chemically or radioactively labeled piece of DNA from a known source. Under carefully controlled conditions the probe will form a hybrid DNA molecule with any DNA on the blot that is highly related (complementary) to it.

The keys to any Southern blot experiment are the selection of the enzymes used to cut the plant DNA and the design of the hybridization probes. Because Ray Sheehy had already mapped all the sites in the Flavr Savr gene construction cleaved by most of the commonly used restriction enzymes, Rick and I had plenty of experimental options. Ray's map gave us the ability to precisely predict the sizes of DNA fragments that should be produced when we cut Flavr Savr tomato DNA with various restriction enzymes and identified them with various probes, which were also designed based on Ray's map.[13] For example, we knew that there were only two recognition sites for the restriction enzyme SphI in Ray's Flavr Savr gene construction and that there were almost exactly 5000 base pairs (bp) of DNA between them. We also knew that the entire Flavr Savr gene and one end of the *kan*[r] gene were present in the 5000 bp of DNA between the two sites. (One of the SphI sites was located within the *kan*[r] gene, and that gene would therefore be cut into two pieces by that enzyme.) If we hybridized a Southern blot of SphI-cut Flavr Savr tomato DNA with a probe comprising the Flavr Savr gene, we should therefore identify that 5000-bp fragment. We could subsequently hybridize that same Southern blot with a probe consisting of

the part of the *kan*^rgene we knew, based on Ray's map, was
also contained between the two SphI recognition sites, and
we should identify the same 5000-bp fragment. And that
was, in fact, exactly what Rick and I demonstrated was the
case for every Flavr Savr tomato candidate we examined.
These results, in conjunction with others, indicated not only
that the Flavr Savr gene construction appeared to have the
same linear map after insertion into a plant as it had before
insertion but also that the *kan*^r and Flavr Savr genes
remained physically linked.

In order to determine the number of Flavr Savr genes,
and especially *kan*^r genes, in our candidate Flavr Savr
tomato plants, Rick and I also hybridized our SphI-cut
tomato DNA with a probe consisting of that part of the *kan*^r
gene we knew was *not* contained on the 5000-bp SphI fragment. This probe was expected to hybridize with a fragment
of each plant's DNA that was part inserted T-DNA (from the
SphI site in the *kan*^r gene to the end of the T-DNA that did
not contain the Flavr Savr gene) and part tomato DNA into
which that T-DNA had been inserted. We had no idea what
size such a "border" DNA fragment would be, since its
length depended on where the closest SphI site in an engineered plant's DNA was in relation to where the T-DNA had
landed in that particular transformed plant. What we did
know was that each border fragment should be unique for
each insertion event, and therefore their number should
equal the number of *kan*^r (and physically adjacent Flavr
Savr) genes in each transformant. And since the kanamycin
germination assays on the 44 plants we were analyzing had
indicated that the Flavr Savr gene construction had been
inserted at only one genetic locus in each plant, we expected
only one border fragment to hybridize for each plant as well.
Instead, we observed two hybridizing fragments for most of

our candidates. Obviously, more than one copy of the Flavr Savr gene construction had been inserted into most of these plants.

Still confident in our germination assay results, we checked the border fragments involving the other end of the T-DNA, the end containing the Flavr Savr gene. If the multiple border fragments we observed on the kan^r side of the inserted DNA were in fact due to multiple independent insertions of the Flavr Savr gene construction, then we would expect to find, for each individual transformed plant, the same number of border fragments on the Flavr Savr gene side of the T-DNA. If, however, we were experiencing a previously documented phenomenon[14] in which *Agrobacterium* -mediated transformation led to the incorporation into one site of multiple copies of the transferred DNA that abutted in a "head-to-head" orientation (kan^r, Flavr Savr gene—Flavr Savr gene, kan^r), we would not expect a "border" fragment at all but, rather, a fragment that represented the junction between adjoining Flavr Savr genes. What's more, for any given choice of restriction enzymes, the length of that junction fragment should be the same in every transformed plant in which the T-DNA had been inserted in the "head-to-head" orientation.

It was at this point, counting genes, that Rick and I eliminated most of the 44 Flavr Savr tomato candidates we had started with. Some of the plants did have multiple insertions at multiple sites in the plant DNA. We suspected that these sites must have been close enough on one plant chromosome that they could not be distinguished by our kanamycin germination (genetic) test. We didn't want to take the chance that those genes would move away from each other in subsequent generations and lessen the intended technical effect of the Flavr Savr gene, and so we eliminated them from further consideration.

Other plants had multiple copies of the Flavr Savr gene construction at one site, some with the T-DNA in a "tail-to-tail" as well as a "head-to-head" orientation. Of these, we eliminated those that appeared to contain four or more kan^r genes. Our border and junction counting method was less reliable for plants with more than three T-DNA regions inserted at one plant DNA site. Not only did we want to be precise about counting our genes, but we also didn't want to overstep the limit of ten genes we'd set in our request of the FDA for an advisory opinion on kan^r. (These first-generation transformed plants were all hemizygous, meaning that the inserted genes were present on only one chromosome of a given chromosome pair. Commercialized Flavr Savr tomato plants would be diploid for the inserted DNA, containing twice the number of kan^r and Flavr Savr genes as their progenitor hemizygous plants.)

We eliminated a few more plants in which truncated parts of the kan^r or Flavr Savr genes had been incorporated. We also lost another couple of good candidates because Matt Kramer found they were off types in the field or didn't produce enough seed to carry forward. When all was said and done, instead of the ten plants that Bill had originally requested, only eight Flavr Savr tomato plants remained that we could dissect molecularly and yet were still commercially viable. But I heard no complaints from Bill when I gave him the news in mid-May. He took those eight plants and ran with them. Characterization of additional intended changes in those eight plants was initiated immediately.

Ray Sheehy documented the expression of kan^r and the Flavr Savr gene by measuring the accumulated RNA that had been transcribed from each, as well as the expected reduction in RNA accumulation from the endogenous PG gene in each plant.[15] Bill (with help from Rick) measured

APH(3')II levels to ensure they were below the 0.1 percent of total protein limit that had been set in the kan^r document and found that levels did not exceed 0.08 percent in any of the eight.[16] Bill and Rick also documented the expected reductions in PG protein levels.[17] In terms of these intended changes in our eight Flavr Savr tomato varieties, we found no surprises.

Safety Investigations

Scrutiny for unintended changes was also set in motion. Demonstrating that there were no unintended changes made to Flavr Savr tomatoes was the major thrust of the "Safety Investigations" section of the tomato advisory opinion document. Keith Redenbaugh was the primary author of that section.

The concern over unintended changes centered on the fact that we had no control over where our genes went into a plant's DNA. They could be integrated harmlessly or smack dab in the middle of an existing gene, disrupting that gene's function. If that disrupted gene was involved in the formation of an important vitamin, for example, levels of that particular vitamin could be dramatically reduced. Of course, unintended changes due to disrupted genes could also occur during the course of traditional breeding, and Keith emphasized that in his introduction to the "Safety Investigations" section. Nevertheless, the FDA wanted specific assurance that the genetic engineering process had not inadvertently caused a decrease in levels of important vitamins or an increase in levels of any tomato toxins. Accordingly, these compounds were measured in the eight example varieties and compared to levels found in control fruit.

Although tomato ranks sixteenth in nutrient concentration, it ranks first among vegetables in the American diet for its contribution of some important nutrients because it is so heavily consumed.[18] Tomatoes are especially high in vitamins C and A, and consequently the FDA was most concerned about the levels of these two vitamins in Flavr Savr tomatoes. A decrease in either of these principal nutrients of 20 percent or greater would warrant reconsideration by the FDA of whether our Flavr Savr tomatoes remained GRAS (generally regarded as safe).[19] Samples of ripe fruit from our eight selected varieties, as well as samples of the appropriate control varieties lacking the Flavr Savr gene, were therefore sent to the National Food Laboratory in Dublin, California, for standard vitamin and mineral measurements.

The good news was that the vitamin and mineral levels measured in the tomato fruit samples tested were within the rather large ranges of published measurements for these compounds in tomato. We concluded, therefore, that no major changes in vitamin composition had occurred in our eight Flavr Savr tomato varieties.[20] There was a caveat in the study, however. Because we had a limited number of tomatoes to work with and many other tests to perform on them, only one measurement per tomato variety had been taken. Statistical analysis of the results had been therefore impossible.

Levels of glycoalkaloids, the only naturally occurring, possibly toxic compounds we knew about in tomato fruit, were also measured in the eight Flavr Savr tomato plants. We knew of no reports of human death or illness due to glycoalkaloid poisoning from eating tomato fruit, but there were numerous such reports about solanine, a glycoalkaloid compound common in potatoes. Solanine is the agent responsible for making green, sprouting potatoes poisonous and is especially dangerous because it cannot be inactivated by cooking. Solanine levels in new potato varieties have been monitored in the United States since the release of a variety called Lenape in the late 1960s that had exceptionally high levels of solanine in otherwise healthy-looking potato tubers.[21]

Levels of glycoalkaloids in tomato fruit, on the other hand, had not been so monitored. In fact, relatively few scientific studies of glycoalkaloid compounds had been carried out in tomatoes at all—at least until the advent of the Flavr Savr tomato. The reason for this lack of interest in tomato glycoalkaloids was simple. The compounds didn't seem to present a problem in tomatoes. Solanine had been detected in tomatoes, but only in green fruit and only at levels much lower than those determined to be safe in potatoes. And the most prevalent glycoalkaloid in tomatoes, tomatine, was estimated to be some 100 times less toxic than solanine.[22] The few studies that had been done indicated that tomatine levels in green fruit decreased dramatically to nearly nonexistent levels as tomatoes ripened. It was conceivable, however, that we could have inactivated, via insertional mutagenesis, a gene responsible for eliminating tomatine during fruit ripening. We deemed it prudent, therefore, to verify that levels of tomatine, as well as solanine and other glycoalkaloid compounds, had not been inadvertently increased in ripe fruit as a result of the genetic engineering process.

Analysis of tomato glycoalkaloids was carried out by John Uhlig, a biochemist with Campbell Soup's Institute for Research and Technology in Davis. On a tight schedule, John modified procedures from the literature and used them to measure tomatine and solanine levels in Flavr Savr tomatoes and non–genetically engineered control fruit. He found no evidence of solanine in any of the fruit samples he analyzed.[23] He also compared measurements from green fruit to those from red fruit for both control and Flavr Savr tomato plants and observed dramatic decreases in tomatine levels like those noted in published accounts. The trend for tomatine levels to drop to extremely low levels in ripe fruit was still apparent in all eight Flavr Savr tomato varieties.[24]

Variability was extensive in John's measurements, however. Tomatine levels could vary twofold among fruit from the same plant. John also found evidence of additional unknown compounds, possibly other glycoalkaloids, in the Flavr Savr and control tomato samples he analyzed. Clearly, Calgene and Campbell Soup's Institute of Research and Technology were teetering on the edge of another new technology, the science of glycoalkaloid determination in tomato fruit. John Uhlig's glycoalkaloid analysis was submitted to the FDA as part of the Flavr Savr tomato advisory opinion, but it was not enough to satisfy the FDA. Calgene would eventually be sent back to the edge of glycoalkaloid technology.

In order to be as conservative and thorough as possible with our safety assessment of the world's first genetically engineered whole food, a general approach for uncovering unintended effects of the genetic engineering process was also undertaken. Called an acute oral toxicity test, it was a standard approach designed to uncover immediate or short-term toxic effects of various chemicals on animals. And although tomato fruit was a far cry from being a single, pure chemical substance like the ones usually tested using this approach, no one at Calgene could think of a better way to reveal unintended changes we could not imagine or, therefore, specifically look for. So Don Emlay sent fruit samples from the eight Flavr Savr tomato plants and five control plants to IIT Research Institute in Chicago, where the studies took place.

The design of the test was simple. Tomatoes were pureed until able to pass through a 16-gauge needle and then force-fed to healthy rats using a syringe. (Rats are often used for these studies since they are convenient and cheap; the applicability of rat data for human safety remains arguable, however.) Each animal received one large dose of tomato puree—approximately the equivalent of a 200-pound person's eating 3 pounds of fresh tomatoes in one sitting—and was then observed under standard lab conditions for 2 weeks. At the end of the observation period the rats were sacrificed and limited autopsies performed on them.

No rats died during the observation period. In fact, mean body weights increased for all rat groups during the study. No gross necropsy lesions or any other problems with the external surfaces of the brains, hearts, lungs, spleens, livers, kidneys, stomachs, gastrointestinal tracts, urinary bladders, adrenals, or gonads of the animals were observed after they were euthanized. The only abnormalities observed, such as "hunched posture," were attributed to the method of force-feeding, not to the tomatoes themselves. IIT Research Institute's conclusion was that there had been no toxicity associated with any of the test article tomatoes or the control tomatoes used in the study.[25] Flavr Savr tomatoes appeared to cause no acutely toxic effects, at least not in rats.

The "Safety Investigations" section of the document also included molecular proof that, once the Flavr Savr and *kan*[r] genes were inserted into a plant's DNA, they would remain at their original point of insertion and behave like naturally occurring endogenous tomato genes. That is, once stably integrated, Flavr Savr and *kan*[r] genes would segregate among progeny plants according to the rules of classic

Mendelian genetics. Since Rick and I had used the genetic germination assay as well as various corroborating DNA tests to determine that all eight selected Flavr Savr tomato plants had T-DNA inserted into only one genetic locus, we expected that T-DNA to be transmitted to the progeny of those eight plants in the same ratio as seeds (progeny) from those plants had germinated on the antibiotic kanamycin, 3:1. We used our original Southern blot–based experimental design, as well as an additional cute molecular trick, to analyze a dozen progeny from each of our eight selected Flavr Savr tomato plants and thereby illustrate the Mendelian behavior of the Flavr Savr gene.

First, we checked each progeny plant to see whether it had the same "border" DNA pattern on a Southern blot (same number and size of DNA fragments) that we had observed in its particular parent plant. Border fragment patterns were established based on the site of insertion and were therefore unique for each parent. If the Flavr Savr and *kan*r genes moved at all from their unique insertion site in a plant's DNA, that plant's border pattern would change. If, on the other hand, the genes remained at their original point of insertion, an offspring inheriting the genes would display the same border pattern as its parent.

We also expected, based on the laws of Mendelian genetics, to see the signature border pattern in only three out of every four progeny we analyzed, with the fourth lacking any border pattern at all. Mendel's laws dictated that, during reproductive cell formation (meiosis), one-quarter of the progeny of our eight Flavr Savr tomato plants would not inherit the chromosome that carried the Flavr Savr gene or genes. DNA isolated from those 25 percent, therefore, would not hybridize with the probes we used for our Southern blot experiments.

In the 92 progeny DNA samples Rick and I analyzed, we found the corresponding parent's border signature in 69 of them (75.0 percent). Not once did the size or number of border DNA fragments differ between an offspring and its parent plant. We also found no hybridization signal on our border Southern blots associated with the other 23 DNA samples (25.0 percent). Our conclusion was that the 23 plants from which those samples had been isolated were the expected quarter of the progeny that had not inherited the Flavr Savr and *kan*[r] genes from their parents. When all was said and done, the Flavr Savr gene had obeyed Mendel's laws of heredity just as well as the gene coding for wrinkled peas had obeyed them for Mendel back in 1866. (On a parent plant–by–parent plant basis, of course, our sample size of 11 or 12 progeny from each was low, and therefore our segregation numbers were not quite as good. In fact, the closest we got to 3:1 "perfection" was 9:2 segregation among the progeny of one plant. Like Mendel, however, we still felt confident of the general phenomenon we were witnessing.)

Then we got more sophisticated. We designed a Southern blot experiment with which we could distinguish progeny plants that had inherited the parental hemizygous (gene present on only one chromosome of a pair) number of Flavr Savr genes and those that had twice that number and were therefore homozygous (gene present on both chromosomes) in terms of our Flavr Savr gene construction. Our method took advantage of the fact that the Flavr Savr gene would hybridize to a PG gene probe just as well as the naturally occurring tomato PG gene would itself. Based on Ray's map of the Flavr Savr gene and a map of the endogenous PG gene published by Don Grierson's group,[26] we carefully selected enzymes that would produce a 967-bp fragment of the Flavr Savr gene and a 1036-bp fragment of the endogenous PG gene. We also designed a probe that would hybridize equally well with both fragments. We then carried out otherwise standard Southern blot experiments with DNA samples isolated from the progeny of each original Flavr Savr tomato plant.

Because there are exactly two copies of the endogenous PG gene in diploid tomato plants (one gene copy on each of two homologous chromosomes), Rick and I had an internal standard for gene copy number in every one of our parent and progeny plants. The intensity of hybridization we observed between our probe and that 1036-bp PG gene fragment represented two gene copies. All we had to do was compare that hybridization signal to the one produced by the PG gene probe hybridizing to the 967-bp Flavr Savr gene fragment, and we could determine the number of Flavr Savr genes in any Flavr Savr tomato plant. For example, if the hybridization intensity of the Flavr Savr gene fragment was the same as that for the PG gene fragment, there were two Flavr Savr genes in that plant. If the intensity of the hybridization signal from the Flavr Savr gene fragment was only half that of the PG gene fragment, then that particular transformed plant had only one copy of the Flavr Savr gene.

The method worked well. For example, from a parent (hemizygous) plant carrying only one copy of the Flavr Savr gene, Rick and I identified six offspring plants in which the hybridization intensity of the Flavr Savr gene fragment was half that of the PG gene fragment, indicating that they too had only one Flavr Savr gene. In another three plants the hybridization signal was equal for the two fragments. These three carried two copies of, and were therefore homozygous for, the Flavr Savr gene. In the other two progeny plants tested, only the 1036-bp endogenous PG gene fragment hybridized at all, indicating that these plants were "null," meaning they had no copies of the Flavr Savr gene. Overall, from among 89 progeny plants analyzed, we identified 47 with the same number (53 percent) and 19 with twice as many (21 percent) Flavr Savr genes as their parents. The same 23 progeny (26 percent) that had lacked border fragment signatures in our previous Southern blot test also lacked Flavr Savr genes in this one.[27] Kanamycin germination assays carried out once the progeny plants produced seed confirmed the hemizygous, homozygous, and null identifications that Rick and I had made months earlier using our Southern blot technique.[28]

I helped Keith incorporate the DNA studies into the "Safety Investigations" section of the document. Keith was not a molecular biologist. He told me he did not want to make any decisions about scientific, especially molecular, issues. Rather, his concern was with meeting the content and format requirements for the federal agencies. (Keith's preoccupation with the details of format, in conjunction with his interest in computer technology, earned him the affectionate title of "Font Master" among the tomato group scientists who worked with him.) So in the final stages of pulling the document together that August, I spelled out the way the DNA results mitigated possible unintended changes that might occur as a result of the genetic engineering process.

Because the *kan*[r] and Flavr Savr genes had been inserted at only one site in the DNA of our Flavr Savr tomatoes, the chances of causing pleiotropic (additional, unexpected) changes by landing in and thereby disrupting an endogenous tomato gene had been minimized. If DNA had been inserted into multiple sites, there would have been multiple chances for gene interruption. Keeping our future options open, we also pointed out that, had insertion occurred at a small number of sites, chances of disrupting a gene would still have been quite low. Once incorporated, kan^r and Flavr Savr genes were immobile in the tomato genome, eliminating any further opportunity for gene disruption via insertional mutagenesis in subsequent generations. Incorporated kan^r and Flavr Savr genes behaved as we expected them to, according to Mendelian predictions.

Bill Hiatt could not have been happier with the DNA results . . . once he had read my report and understood just what Rick and I had done anyway. We now knew the precise number of Flavr Savr and kan^r genes in (potentially) commercially viable tomato plants and could assure the FDA that we were well under the limit we had set for ourselves in our previous request for an opinion on kan^r. In addition, we had counted genes using two independent and corroborating methods, methods that showcased the precision of ag biotechnology quite nicely.

The method Rick and I had established for discerning the genotype of progeny plants—hemizygous versus homozygous for the inserted genes—was also a boon for product development. For breeding purposes, the homozygotes were the only plants Matt Kramer was interested in. We could now save time and greenhouse space by selecting them from among their many hemizygous siblings as young plants instead of waiting for seed with which to carry out kanamycin germination assays.

The Public Comments

Our results were timely in another way as well. The FDA had requested comments from the public on Calgene's previous request for an advisory opinion on the use of kan^r in genetically engineered plants. The period for public comment ended on July 30, 1991, as we were putting the finishing touches on the Flavr Savr tomato document. Copies of the comments arrived at Calgene on August 7. We barely had time to read them, much less address them experimentally, prior to August 12, our intended filing date for the second document. So Bill was particularly pleased that our DNA analysis already addressed one of the major concerns of one of Calgene's biggest dissenters. The National Wildlife Federation was concerned about Calgene's failure in the previous document to demonstrate its ability to count inserted genes. Margaret Mellon had called it a serious omission. Determining the number of copies of a foreign gene incorporated into a recipient cell was, she wrote, one of the "many aspects of modern gene transfer technology that lacks precision." Furthermore, "if the company cannot assure this relatively low level of gene incorporation [ten *kan*[r] genes or less], many of its safety arguments are undermined."[29] The thorough investigation that Rick Sanders and I had done to determine the numbers of inserted genes in our Flavr Savr tomatoes addressed Mellon's concern in spades. Bill beamed as he congratulated me over this little victory.

I skimmed over the careful critiques prepared by the National Wildlife Federation and Environmental Defense enough to note that many of both groups' concerns were with the environmental assessment Calgene had filed as part of its *kan*[r] request. I was not surprised. The National Wildlife Federation, in particular, even suggested experiments Calgene could carry out as a way of addressing specific issues, such as gene flow from engineered canola, that concerned them. The public comments convinced me that the decision to address environmental issues associated with commercial growing of Flavr Savr tomatoes to the U.S.

Department of Agriculture (in filing for an exemption from plant pest status) instead of the FDA had been a good one. A Flavr Savr tomato environmental assessment would be all the better for the postponement. An environmental assessment per se was not included with the Flavr Savr tomato request for an advisory opinion filed with the FDA.

Besides comments from the environmental groups and five individual consumers, most of the comments submitted to the FDA had been from university professors using the techniques of genetic engineering in their own research and representatives of various agricultural industries with their own vested interests in the new technology. Consequently, a cursory tally of the pro versus con comments among the total of 43 submitted indicated that the overall public response was extremely positive—at least from Calgene's point of view. Accordingly, in a thank you letter he sent to the FDA respondents, Roger Salquist described it as an "overwhelmingly positive response from the academic and business communities."

It was in this jubilant mood that Calgene filed the Flavr Savr document a few days later. Don Emlay was positively gleeful. He was so sure of public and FDA support for our tomato that he was willing to wager the Flavr Savr tomato document marked the end of our data submissions to the FDA. I took him up on it. He got more specific. "If the FDA required any more information from us," he claimed, "it would be minimal, no more than two weeks' worth of experiments at most." How I wish that I had lost that bet!

The Celebration

In-house Calgene celebrations marking the filing of the "Request for Advisory Opinion, Flavr Savr Tomato: Status as Food" were similar to those held when the *kan*[r] document was filed, although considerably more private. Those of us who worked on the project again received a nominal bonus and a T-shirt (tomato-red), but this time Bill presented the awards individually in his office. The seclusion was an attempt to avoid the grumbling that had occurred among the rest of the Calgene science staff after the *kan*[r] document bonuses had been bestowed. (As one might expect at a small company like Calgene, however, the effectiveness of that strategy was debatable.)

There were no Plexiglas "tombstones" containing miniaturized versions of the tomato document, a gesture I took as a sign from Bill that our effort had been science- and not business-driven. He did, however, throw a party at his home for the entire tomato regulatory crew. Two Calgene parties at Bill Hiatt's house within 9 months. Astounding!

Publicly, Calgene put out a press release.[30] It quoted Roger Salquist as saying, "The tomato's status as a food is unchanged by the insertion of the Flavr Savr gene" and mentioned his gratification with the "overwhelming positive response" the company had inferred from the comments the FDA had received concerning Calgene's *kan*[r] submission. But that press release did not simply celebrate the regulatory victories of the moment. For one thing, it marked the first time the moniker "FLAVR SAVR" (originally spelled in all caps) was used in public. Roger had become upset as he reviewed the (supposedly) final draft of the FDA tomato advisory opinion document in which a more scientific-sounding term had been used: "There is no god-damned way we're going to use that god-damned name for our product." In a flash of what Keith Redenbaugh and other Calgeners came to refer to as "Roger's usual brilliance," Roger had come up with "FLAVR SAVR" on an airplane about a year earlier when he realized that even he could barely spit out "antisense polygalacturonase." (Of course, he seldom tried and referred to the enzyme as "polygal" instead.) Keith hurriedly substituted "FLAVR SAVR" for the commercial version of the gene throughout the document and filed it with the FDA.

That press release additionally announced that Calgene had "signed an agreement with Campbell Soup Company for the exclusive license to the use of the [Flavr Savr] gene for fresh market tomatoes in North America." Considerable space was also dedicated to reiterating the potential consumer benefits of a tomato that could be "harvested vine-ripe, with enhanced flavor and extended shelf life." The transition from the regulatory phase of commercializing the Flavr Savr tomato (at least the proactive part of it) and the marketing business phase of that effort was underway. Calgene was getting down to the business of selling genetically engineered tomatoes. Life as a tomato scientist at Calgene was about to change forever.

CHAPTER FIVE

Transition from a Science-Driven to a Business-Driven Enterprise

During Calgene's transition to business mode, life changed faster for some tomato scientists than for others. I held out for as long as possible. I spent the first couple of months after we filed the Flavr Savr tomato advisory opinion document with the FDA doing science, non–regulatory related science.

Calgene had just signed a 1-year contract with Campbell Soup Company to test a couple of other potentially promising genes in tomato, and Bill Hiatt, who was already otherwise occupied with the business transition, had put me in charge of managing it. That kept me busy at the lab bench and in my office. Rick Sanders and I also wrote a paper describing our gene-counting method and submitted it for publication despite a comment Bill made about keeping it a "trade secret." I savored what turned out to be the last vestiges of academic-style science for Calgene's tomato group.

My scientific interlude didn't last long. By October 1991, the tomato business that had been brewing in earnest since the first of the year made its initial impact on Calgene's science staff. Roger Salquist held another "reorganizational meeting." In preparation for selling Flavr Savr tomatoes, he said, a marketing arm of the company, named Calgene Fresh, Inc., was being established. Calgene Fresh would be a fully owned subsidiary of Calgene but have a separate management team and keep separate books. There were no plans to sell off this subsidiary, we were assured, but in order to facilitate separate bookkeeping the tomato scientists would be consolidated into one lab.

Up until then Calgene's scientists had been randomly situated throughout the company's labs, scientists who worked on tomato mixed together with those who worked on cotton or canola. Integration had advantages. Although we had somewhat different methods of isolating DNA, RNA, and proteins from the different crop plants we worked with, once we had those molecules in hand we carried out very similar operations with them. We shared ideas, techniques, reagents, successes, and failures. Close proximity made it easy to learn from and teach each other.

From our point of view, easier bookkeeping was an inadequate reason to segregate us. Our research efficiency would inevitably be reduced, especially during the months it would take us to get used to new lab surroundings. What's more, we didn't buy the idea that there was no plan to sell off the new tomato subsidiary. Rumors that Calgene Fresh would issue its own stock were already flying. Most of Calgene's scientists, especially those not associated with tomato research and therefore unlikely to receive any of the subsidiary's stock, were unhappy with the whole reorganization idea.

We'd been unhappy with prior reorganizations too, but this one presented a new twist. Previously, our unhappiness had been largely directed at the executive staff. From our point of view, they made the mistakes, but we had to pay for them, usually through layoffs. But this time the reorganization of the company around Calgene Fresh planted the potential for strife within the scientific ranks. Wouldn't the tomato group get preferential treatment? Better lab space? What about those rumored stock options? The usually united scientific front at Calgene started to crack.

The situation only got worse after Roger handed over control of Calgene Fresh to Thomas L. Churchwell. Tom had been a member of Calgene's board of directors since 1987. His expertise was sales. As a vice president with NutraSweet® Company from 1983 to 1987, he'd sold the first NutraSweet contracts to Coca-Cola™ and Pepsi™. When word hit the labs that Tom thought the Flavr Savr tomato, with related regulatory documents already filed, an issued patent on the gene, and especially its consumer appeal, was the "best business opportunity" he'd come across since NutraSweet, we found his opinion reassuring. Apparently sold on the Flavr Savr tomato concept himself, Tom eagerly signed on as president and CEO of Calgene Fresh in order to market it. At first blush, he seemed a great choice for the job.

It didn't take long, however, before Calgene's choice for tomato commander was second-guessed. Despite the fact that the major centers for growing fresh tomatoes in the United States are the Central Valley of California, where Calgene, Inc., was located, and parts of Florida, Tom decided that the central offices of Calgene Fresh, Inc., would be in Evanston, Illinois. "We wanted to be as far away from good tomatoes as possible," he said about the move later.[1]

The decision made little sense to anyone. After Tom first presented the idea to the board of directors, Howard D. Palefsky, president and chief executive officer of Collagen Corporation, questioned him about the move to Evanston. Howard, who likely had more business operations experience than anyone else in the room, found Tom's response unconvincing, and although he had established, during his 5-year tenure on Calgene's board, a reputation for registering his disapproval subtly, this time he made an exception. Howard Palefsky thought moving Calgene Fresh to Evanston, Illinois, was a bad idea, and he let Roger Salquist know it.

But Roger was caught between a rock and a hard place over the Evanston situation. He believed in delegating and had agreed to take a hands-off approach to Tom and Calgene Fresh. For his part, Tom was well aware of Roger's dominant personality. Roger had been a Navy submarine officer, among the first group of history and political science majors (out of Harvard, Stanford, and Rensselaer) to have passed muster with Admiral Rickover, and the military disciplinarian often came out in him. Tom likely figured that a separation of 2000 miles might be necessary to ensure minimal surveillance by his boss. (Another obvious reason for moving the company to Evanston was that Tom was living and working there at the time.) Roger was stuck. His choices were to overrule Tom's decision, likely losing Tom and his sales strengths in the process, or to stay out of it. He kept to his hands-off agreement.

Consequently, the stage was set. Calgene, Inc., was staking its future on Calgene Fresh and the Flavr Savr tomato. And the undisputed leader of Calgene Fresh was Tom Churchwell. The Flavr Savr tomato would be launched according to his flight plan.

The New Boss

Tom held a "meet the new boss" gathering for the tomato scientists that December. We were relieved to learn that he wasn't taking his science staff with him to Chicago. Research and Development for Calgene Fresh would remain in California, under the same roof as its parent company, although with its own, newly constructed private entrance. But Tom didn't talk about Calgene Fresh's impending administrative move at that meeting.

Instead, he spoke at length about the new corporate culture we, the employees of Calgene Fresh, would develop. He sounded both sincere and excited as he revealed his vision for the venture ahead. An important part of our new corporate culture, he told us, would be open communication among all the employees at Calgene Fresh, from the dishwashers right up to him. In fact, it was not only our right but also our responsibility to speak out if we felt the company was going in the wrong direction. Tom's description of the corporate culture he envisioned for Calgene Fresh sounded incredibly ideal, a worker bee's dream.

Many of us, however, felt that the corporate culture of Calgene, Inc., already allowed for the kind of teamwork and open communication Tom was describing. Roger Salquist himself had an "open-door" policy. Granted, there was always a chance that Roger might yell you out of his office, but that was usually when he didn't like your idea, not because he wouldn't listen to it. Was it really necessary to splinter us off, both physically and culturally, from the rest of the company? Matt Kramer took advantage of the communication policy Tom was espousing to question this aspect of his corporate game plan.

"Why do we need a new corporate culture?" he asked, "What's wrong with Calgene's corporate culture, the one we already have?"

Clearly, this was not the kind of response Tom had expected his pep talk to elicit. It was uncomfortably obvious from his body language and the tone of his voice that he was not at all pleased with the results of this first experiment in cultivating Calgene Fresh's corporate culture. Matt believed he had gotten himself into the doghouse with Tom, and Calgene Fresh had not even been officially christened yet.

Tom's reply was notable, not for its content, but for the manner of his speech. It was our first taste of the new boss's use of circumlocution. It was a brand-new experience for me, one I found very disconcerting and never got used to.

I'd been listening to executives at Calgene for nearly 4 years by then. I could deal with receiving information that was less than the truth, the whole truth, and nothing but the truth. Roger was occasionally blatant about it. "You know you're in trouble when you start believing your own press releases," he once declared during an all-employee meeting. True or not, at least I always understood what the members of Calgene's management staff were trying to say.

Things were different with Tom. Sometimes I simply couldn't follow his logic at all. Figuring out what he was trying to say felt like "untossing a salad."[2] And asking him for clarification, at least for me, usually just made the situation worse. In *The Elements of Style,* E. B. White noted that a "good many of the special words of business seem designed more to express the user's dreams than his precise meaning."[3] In my opinion, that description fit Tom Churchwell to a T.

Others involved in business, I've since learned, successfully practice what Nobel laureate economist Robert M.

Solow of MIT calls the "art of meaningless verbiage."[4] Alan Greenspan, chairman of the Federal Reserve, for example, has been called a master of it.[5] Circumlocution, so it seems, is therefore not necessarily viewed negatively in the world of business. Mr. Greenspan, in fact, has told audiences to expect it from him. "I guess I should warn you, if I turn out to be particularly clear, you've probably misunderstood what I've said."[6]

But effective communication, and lots of it, was the main message Tom did manage to convey during that first meeting with his Calgene Fresh scientists. How could Tom, with his manner of speaking, facilitate effective communication? And what about sending half of the new company halfway across the country? That certainly wasn't going to make communicating among Calgene Fresh employees any easier. It all sounded like a recipe for communicational failure to me. As far as I could tell, Calgene Fresh was off to a perilous start.

An Organization All about Quality

On February 12, 1992, Bill Hiatt handed me, in the privacy of his office, an official letter of transfer from Calgene, Inc., to Calgene Fresh. Bill, whose title was now vice president of research and development for the subsidiary company, was very formal about it. He shook my hand. "I hope you're ambitious," he said very seriously. Since my most logical next career move would have been to pursue his job, I interpreted the sentiment as forewarning of the sheer amount of work that lay ahead of us. He closed the ceremony by asking me to keep the number of stock options I'd been granted to myself.

The letter was from Tom Churchwell. Besides delineating the number, exercise price, and vesting structure of the "pre-

IPO" (I viewed them as virtual) stock options I had been granted, it spelled out Tom's idea of the road ahead. "We have the opportunity to create a truly unique environment in which unparalleled corporate accomplishments will be achieved without cost to family and personal goals. I look forward to working with you to make Calfresh's [the in-house shorthand name for the subsidiary] dreams come true."

We'd been busy defining Calgene Fresh's dreams for a while by the time we got our official transfer letters. Since New Year's 1992, the actual birth date of the subsidiary company, all 19 "founding members" (consisting of Tom, 3 vice presidents, 1 accountant, 2 secretaries, and 12 scientists) had gotten together regularly in an effort to put together our "Organizational Vision." Although it would be another 6 months before we had hashed everything out, from our "Core Values and Beliefs" through our "Vivid Description," it was clear from the outset that Tom's clearest vision for this company had everything to do with "quality." We were given books on the subject. We discussed and defined it, in general and as it related specifically to tomatoes regularly at these (and all other, for that matter) Calgene Fresh meetings.

These organizational vision meetings were facilitated by Stephen C. Benoit. Steve had come to Calgene in October 1987 by way of PFC Group, Inc., where he had worked for 6 years. He, like Don Emlay before him, had been one of those Calgene directors (in his case of planning and financial analysis) who initially caused the science staff to wonder just what his job entailed. His involvement in acquisitions and mergers soon became all too apparent, however, when news of an impending deal to purchase a flower farm in Oregon trickled down to the labs. Luckily, a research associate responded by relaying information he had about toxic water readings at that particular farm to Steve and Dan Wagster. They killed the deal immediately. Years later, over beers at a local brew-pub, Steve and Dan could finally laugh about the near disaster.

Besides his prowess on the Calgene softball team, Steve was most widely known among the science staff for his promotion to vice president of strategic planning and analysis in September 1990. Roger Salquist made the announcement at an all-employee meeting. He explained that, while Steve's duties would not necessarily change, he needed the title of vice president in order to carry out those duties properly. While you had to admire Roger's honesty, this example of his promotion methods did nothing to bridge the gap between the business and science cultures at Calgene.

With help from Matt Kramer, Steve had put together a fresh-market tomato strategic marketing plan and presented it to Calgene's executive staff in January 1991 and its board of directors the following August. That plan supplied the impetus to initiate Calgene Fresh. Steve was made vice president of marketing for the new subsidiary. Along with Tom Churchwell, Bill Hiatt, and Dan Wagster, as vice president of operations, Steve Benoit rounded out Calgene Fresh's initial management team.

A key to the plan was that Calgene Fresh would charge a premium for its tomatoes of two or three times the price of "gassed green" fruit because of their higher quality. It was a novel idea, since there were essentially no premium tomatoes on the market at the time. Tomatoes that "looked red but ate green" were the consumers' only choice. The Flavr Savr gene was expected to provide a certain amount of the higher quality required, but innovative picking, packing, and distribution systems would also be necessary to deliver premium tomatoes to market quickly and carefully enough to preserve whatever added quality they had. Quality was the key.

When, in the course of our organizational vision meetings we got down to our tangible need, or not, to form partnerships with companies that might know more about tomato quality than we did, I suggested that we might seek counsel from our friends at Campbell Soup. Campbell had, in the late 1980s, tried its hand at selling branded premium-quality fresh-market tomatoes. Steve and Dan quickly reminded me that the Campbell foray into fresh produce had been a failure, and by June 1990 Campbell had given up the effort. "Even so," I persisted, "perhaps we could learn something from their mistakes."

"When the Calgene board of directors invited me to form Calgene Fresh," Tom Churchwell responded, "they said, 'The good news is that you know nothing about the tomato business so you won't fall into the old traps that they have. The bad news is that you know nothing about the tomato business and you'll fall into traps that they know how to avoid.'"

For a moment I was afraid Tom meant we would go it alone. I was greatly relieved to learn otherwise. With only limited tomato-growing experience (consisting at the time of seven small-scale field trials) and essentially no experience in packing and shipping tomatoes for market, Calgene Fresh management needed to form a partnership with an established grower-shipper company. And they knew it. But, instead of turning to Campbell Soup, known for processing tomatoes (to the tune of 315 million cans of soup consumed annually in North America[7]), not for fresh tomatoes, they wanted a partner both committed to "farm stand–fresh" quality and willing and able to supply Calgene Fresh with tons of tomatoes. Dan Wagster started scouring the country for one.

Building a brand name for our premium tomato was another key to Steve's marketing plan. Branding Calgene Fresh tomatoes would encourage repeat business from people who recognized their "farm stand–fresh" quality. Tom had already declared that Calgene Fresh would sell its first top-of-the-line, premium tomato by the end of its first calendar quarter. We needed to define the product specs for those high-quality tomatoes and give them a catchy, market-friendly brand name. Consequently, the founding members also spent hours during those first all-organizational meetings collectively brainstorming an appropriate trademark. We filled wall-length white boards with possibilities. We submitted them by e-mail.

But more than simply for coming up with a product name or defining quality, these all-employee meetings were designed to build an effective interface between Calgene Fresh's R&D staff and its business managers. We had been bombarded with articles and other materials on that subject since Calgene Fresh's birth, many of them related to a book called *Third Generation R&D: Managing the Link to Corporate Strategy*,[8] written by three employees of Arthur D. Little, Inc., the international management and technology consulting firm. The whole process was part of Tom's vision of drawing the scientists into his business and creating that common language and culture he'd so earnestly preached at our first meeting. Theoretically at least, we were putting together an interdisciplinary team.

FinelyRipe was my personal favorite for trademark. The team consensus, however, seemed to be that JustRipe was better. And JustRipe was in fact the trademark Tom used in the first-year operating plan he submitted to Roger Salquist at the end of January.

But Steve Benoit decided otherwise. He sent around an e-mail one morning informing everyone that our top-of-the-line tomatoes would be called MacGregor's®. His epiphany had occurred, he explained, while reading to his kids the night before. The name would stand for farm stand–fresh, great-tasting tomatoes from the good old days. And if it brought to mind a familiar childhood story, so much the better. From the sound of Steve's note, it was a done deal. Calgene Fresh's premium product had been named.

Most of us didn't like the name. Tom told the board of directors that "none of us is satisfied with the [trade]mark." But Tom supported Steve's unilateral decision anyway. "Nobody liked the name NutraSweet[R] at first either," he told us. He didn't seem concerned about the circumvention of process that had led to the adoption of the unpopular name. Steve went ahead and ordered packaging and supplies stamped with the MacGregor's trademark.

The MacGregor's tomatoes Calgene Fresh would sell as part of its test—or what Tom preferred to call a learning—market were of necessity conventionally produced. Genetically engineered MacGregor's quality tomatoes were not yet ready for market in early 1992. There weren't enough seeds from the Flavr Savr tomato varieties in the product pipeline to produce the number of tomatoes necessary for even a small entry into the market. And we weren't even sure that those genetically engineered varieties were the ones we wanted to go to market with. In case they weren't, the Flavr Savr gene was being used to transform as many additional and, it was hoped, more commercially viable tomato varieties as Calgene Fresh's small transformation-regeneration staff could handle.

But the biggest reason the first MacGregor's tomatoes were not genetically engineered was that Calgene hadn't yet received the approval it had sought from the FDA to sell tomatoes containing *kan*^r and the Flavr Savr gene. Until such approval was obtained, Calgene Fresh's business team would conduct its test market with conventionally grown tomatoes. In the meantime, the principal priority of the company's R&D staff was ensuring that FDA approval did occur, it was hoped by early 1993.

All indications were that the FDA would eventually approve the use of *kan*^r in our products. However, the use of an antibiotic-resistance gene was high on the list of objections that opponents of crop genetic engineering had at the time. A few of the otherwise supportive letters submitted during the open comment period for the *kan*^r advisory opinion document had, in fact, suggested that Calgene eliminate it or find an alternative selectable marker gene. So, in the interest of public acceptability, a significant effort was initiated at Calgene Fresh to evaluate the feasibility of doing without *kan*^r. We couldn't, however, count on that effort's panning out. Answering questions the FDA had already started asking related to both the *kan*^r and the Flavr Savr tomato advisory opinion documents Calgene had submitted therefore became top priority for Calgene Fresh R&D.

At first Bill Hiatt toyed with the idea of preparing another, relatively large, but all-inclusive document for the FDA and filing it with the agency in early 1992. It would have provided additional data on Flavr Savr tomato varieties included in the tomato advisory opinion document filed the previous August, as well as quality-control data on new Flavr Savr tomato varieties that had entered Calgene Fresh's product pipeline since then. But by the end of February that plan was scrapped. I believe an experimental "discovery" of sorts that Rick Sanders and I had made contributed to Bill's change of heart.

A False Start
for Calgene Fresh R&D

To further substantiate our claim that the Flavr Savr gene, once inserted into a plant's genome, was as stable as naturally occurring Mendelian genes, Rick and I were carrying out follow-up experiments. For the FDA tomato advisory opinion document, we had examined only the original transgenic Flavr Savr tomato plants and their offspring. Those offspring had since produced offspring that had also produced offspring. We therefore had our first opportunity to check these grand- and great-grand-progeny for stability of the Flavr Savr gene.

Leaves of the grand-progeny of several of the original plants had been collected for us from a field trial conducted in Manteca, California. We isolated DNA from the leaves and used it to carry out what was by now our standard Southern blot analysis. Looking at the results, we were surprised to find that the pattern of border fragments in the grand-offspring of one plant, number 502-2021, was different than the pattern in the original number 502-2021 parent plant. But before Rick and I thought to worry about genetic instability, we both noticed something else. The unexpected border fragment pattern looked familiar. We thought we'd seen that same pattern before.

We sifted through our old Southern blots, and, sure enough, the surprising pattern was identical to the border fragment pattern of a completely different Flavr Savr tomato plant we had previously analyzed. We didn't believe for a moment, though, that the Flavr Savr gene had jumped out of its original site in plant 502-2021's DNA and into this other site. The chances of that happening, that the gene would jump into what appeared to be the same site that the Flavr

Savr gene occupied in the other independently produced transgenic plant, seemed just too improbable. There had to be another explanation.

Our theory was that leaf tissue had been mistakenly collected from the wrong plants in the field. We looked over a copy of the map of the Manteca field trial. My bet was that seeds of the two transgenic plants had been sown next to one another, and, sure enough, they had been. Ours was a case of mistaken identity, not roaming transgenes.

While disappointed in the setback for our data gathering time-wise, Rick and I couldn't help but feel a little excited about our sleuthing capabilities. The border fragment patterns were like fingerprints that identified our plants. We could utilize them for quality-control measures.

Bill Hiatt didn't see it that way at all. In fact, he saw it as just the opposite, a complete failure of quality control. And that failure was amplified by the fact that plant 502-2021 was critical to both Calgene Fresh's regulatory and its product development plans. The hybrid Flavr Savr tomato variety deemed most commercially viable at the time was produced using plant 502-2021 (or, rather, its progeny) as the female inbred parent. Of the eight Flavr Savr tomato varieties included in the FDA advisory opinion document, plant 502-2021 was the only one that could be so used in the critical genetic cross. If across-the-board approval for any and all Flavr Savr tomato varieties was not received from the FDA and only those varieties for which we had provided complete data were given the go-ahead, plant 502-2021 might represent our only shot at a product. "The _ _ _ _'en rat is out of the cage," Bill raged when I gave him the news, "this experiment is out of control." Soon after the mislabeling episode, Bill dropped his plan for submitting a large data package to the FDA.

A Better Start for Business

A few weeks later, however, even Bill had to admit that things at Calgene Fresh were looking up. That's when Roger Salquist and Jerry Caulder, chairman of Mycogen Corp., which developed the first approved genetically engineered pesticide, took charge of the lobbying effort of the Biotechnology Industry Organization (BIO). Although pharmaceutical biotech companies made up the vast majority of BIO's members, Salquist and Caulder managed to swing the organization's efforts over to the agricultural side of the industry, described by some as Wall Street's "ugly ducklings of biotech."[9]

Roger met personally with Vice President Dan Quail, head of the Bush administration's Competitiveness Council. He wanted to break what he felt was a regulatory "logjam" caused in part by an "internal administration battle over how to regulate ag-biotech products."[10]

The effort worked. In late February 1992, President Bush officially endorsed genetic engineering of crop plants and "set plans to streamline the regulatory process"[11] for getting them to market. It was not only good news but great publicity. Following the White House announcement, Calgene and its Flavr Savr tomato were the cover story of *Business Week*[12] and mentioned in *Newsweek*.[13] Backing from the Oval Office certainly had its advantages. (Calgeners were thankful that the first product in our pipeline wasn't genetically engineered broccoli.)

Roger Salquist and Tom Churchwell sold 2 million new shares of Calgene stock on Wall Street the following month. The offering raised $22.5 million. Tom said Calgene's tomato science, represented by what he called the red book (a CRC Press copy of the Flavr Savr tomato advisory opinion document submitted to FDA[14]), had made a big hit with investors. By the ides of March, Calgene Fresh seemed on a roll.

To round out the month, Calgene Fresh sold its first tomato on what most people would call April Fool's Day, 1992. To ensure fulfillment of his prophecy that the company would be selling products by the end of its first calendar quarter, Tom Churchwell called it March 32 instead. Whatever you called it, that was a happy day at Calgene Fresh.

Later that April, U.S. Patent Number 5,107,065, "Anti-Sense Regulation of Gene Expression in Plant Cells,"[15] was issued. It covered not just the Flavr Savr gene specifically but the broad use of antisense technology to shut off any gene in any plant cell. Calgene's patent position could not have looked better.

Even culturally, we seemed to be making great strides. We'd worked hard and long at our frequent Calgene Fresh–wide meetings to establish core values, distill a mission statement, and define a vivid description of the organization. It had been tedious going. By the second week in April we still hadn't agreed on our company purpose. One Friday afternoon, 3 weeks overdue with my first child and planning not to come in on Monday no matter what, I sat alone in my office and wrote one: "Calgene Fresh, Inc.'s purpose is to improve the quality of human life through the development and utilization of innovative agricultural technologies." I e-mailed it to Steve Benoit.

Steve loved it. At the company TGIF gathering that afternoon he said that, but for minor adjustments, the purpose I had penned was final. Initially I felt a twinge of guilt for circumventing the consensus process myself, but it evaporated when it turned out the rest of our Calgene Fresh colleagues liked it too.

The final version of Calgene Fresh's statement of purpose: "Using our creativity and innovation, our purpose is to improve the quality of human life through the development, production, and delivery of food that is of the highest quality, flavor, and freshness," was one I could live with. As a bonus, I had "effectively interfaced" with Steve and, I hoped, through him, with the rest of Calgene Fresh's management. I departed for maternity leave on a high note.

The FDA's Policy toward the Ag-Biotech Industry

On May 29, 1992, the FDA published its "Statement of Policy: Foods Derived from New Plant Varieties,"[16] in the *Federal Register*. This "regulatory roadmap"[17] sounded pretty familiar to those of us at Calgene who had been dealing with the agency over at least the previous 18 months. As we had done in our requests of the agency, the FDA compared and contrasted crop genetic engineering with traditional breeding. The "genetic modification techniques" comprising these two disciplines were described as constituting a "continuum."[18] Based in part on the idea of this continuum, the FDA reaffirmed its intention "to regulate foods produced by new methods, such as recombinant DNA techniques, within the existing statutory and regulatory framework."[19] This was a reaffirmation of a policy originally described, in conjunction with the Office of Science and Technology Policy in the Executive Office of the President, in the *Federal Register* of June 26, 1986.

The new statement went beyond the earlier policy, however, by outlining specific types of changes associated with "new plant varieties" that "may require evaluation to assure food safety."[20] These were the issues that we at Calgene were only too aware of, many of which we were still grappling with at the time. Bill Hiatt and Keith Redenbaugh had already prepared responses to the FDA's concerns about the potential toxicity or allergenicity of an introduced protein [in our case, APH(3')II] and whether use of a selectable marker gene conferring antibiotic resistance (in our case, *kan*^{r}) could reduce the therapeutic efficacy of the antibiotic eaten simultaneously with a plant containing such a gene. Likewise, Keith was already aware that the FDA would "to the extent possible, . . . rely on APHIS [Animal and Plant Health Inspection Service] NEPA [National Environmental Policy Act] reviews"[21] for any environmental considerations related to these new plant varieties. In fact, Keith had Calgene's petition requesting removal of the Flavr Savr tomato from USDA APHIS's "plant pest status" ready to go. Keith and Bill submitted their additional information to the appropriate government agencies within days of the release of the FDA's policy. [Additional data supporting Calgene's conclusion that APH(3')II would not compromise antibiotic therapy were submitted to the FDA October 30, 1992.]

The new FDA policy also suggested that "unintentional" changes that resulted from genetic modification or changes that could result in increases in "known toxicants" or decreases in "nutrients" normally found in a food would have to be evaluated. Calgene had, of course, measured glycoalkaloids and vitamins A and C in its Flavr Savr tomato varieties to make sure that unintentional changes in the levels of these compounds had not occurred during the genetic engineering process. And the company's plan was to continue to make these measurements, as part of Calgene Fresh's overall quality assurance program, on new Flavr Savr tomato varieties destined for market.

All things considered, it looked as though Calgene was right on track in terms of the FDA's policy . . . except perhaps for one item: pleiotropic effects. The agency mentioned the possibility of these unknown unintentional changes in its policy statement, but only briefly. The agency implied that plant breeders could deal with the issue "using well-established practices,"[22] the same argument Calgene had made in its previous submissions to the agency. Although the FDA stated that "feeding studies or other toxicological tests may be warranted when . . . safety concerns . . . cannot be resolved by analytical methods," it also indicated that "feeding studies on whole foods have limited sensitivity because of the inability to administer exaggerated doses."[23] In an article published the month following the release of the FDA's policy, David Kessler, FDA commissioner, and fellow authors expanded on this dosage difficulty, stating that "traditional animal toxicology tests [are] designed to assess the safety of single chemicals," not "whole foods, which are complex mixtures."[24] It seemed clear that the FDA was downplaying animal feeding studies as a way to address possible pleiotropic effects.

But just 6 months earlier, the agency had encouraged Calgene to conduct a 28-day study of rats fed Flavr Savr tomatoes. The consensus had been that everyone would breathe a little easier on the issue of pleiotropic effects if such a "wholesomeness study," as FDA's Jim Maryanski referred to it, was carried out. By the time the FDA's statement of policy was published, one wholesomeness study of a Flavr Savr tomato variety had already been concluded, and another of two more recent tomato product candidates was just getting under way. Despite the apparent change of heart on the issue by the FDA, Calgene had already made a date with destiny on the issue of rat feeding studies.

FDA's Policy and Public Response

Calgene and the Flavr Savr tomato were specifically mentioned in the FDA's "Statement of Policy: Foods Derived from New Plant Varieties." The agency explained that, because Calgene had approached the FDA before its policy was finalized, it had advised the company to submit its information as a request for an advisory opinion. (The agency made it clear, however, that future requests from producers of new plant varieties developed using recombinant DNA techniques "should be made consistent with the principles outlined in"[25] its statement of policy.) The FDA had so advised Calgene in order "to utilize an evaluation process that is open to public comment and permits the agency to make its decision known to the public." The FDA went on to direct policy readers to a filing announcement for Calgene's Flavr Savr tomato advisory opinion request and its accompanying call for written public comments located elsewhere in the same issue of the *Federal Register.*

The FDA asked for public input not only on Calgene's advisory opinion requests but also on the agency's statement of policy. In many places throughout the statement's text, the agency called for both general comments and specific comments on issues such as labeling. Letters from the public regarding the FDA's policy, Calgene's request of the FDA for an advisory opinion on the Flavr Savr tomato's status as food, and the petition Keith filed with the USDA to deregulate Calgene's tomato were all due the end of August.

The public's response to Calgene's Flavr Savr tomato–related requests of both the USDA and the FDA was, as had been the case with its selectable marker gene request, minimal. The USDA received only two dozen letters, the FDA even fewer. Most of the comments, from university professors and organizations connected to the food industry, supported allowing the Flavr Savr tomato to be grown and marketed like any other tomato. Predictably, however, Environmental Defense, the National Wildlife Federation, and Jeremy Rifkin's Foundation on Economic Trends opposed what they saw as a lack of oversight of the ag-biotech industry by the federal agencies. Representatives from each organization argued that Calgene's tomato contained "food additives" and therefore required premarket safety testing under the auspices of a food additive petition. Jeremy Rifkin specifically warned that, as a test case, the "FDA's posture on this first whole food product of recombinant DNA technology will set the stage for the scores of other transgenic crops that are currently under development."[26]

On the "test case" issue, at least, Rifkin and his foundation were absolutely correct. A sample public comment letter to the FDA written at Calgene put it this way: "The commercialization of the Flavr Savr tomato will demonstrate to the public that products of biotechnology are safe and the FDA's science-based approach to the safety evaluation of these products does work." But the fact that Rifkin was right didn't keep Roger Salquist from referring to him as an "idiot" who "isn't taken seriously by anyone who can spell their own name."[27] And, despite Rifkin, who also found the Flavr Savr tomato without "any redeeming value" and predicted it would be "dead on arrival,"[28] the environmental groups, and the approximately 5000 public comments the FDA had received that were related specifically to the agency's statement of policy—many complaining about lack of oversight and calling for labeling—Calgene (Calgene Fresh) scored this second round in the fight for public acceptance as a win. In fact, Calgene Fresh was so confident that federal approvals for its Flavr

Savr tomato were imminent that, not one, but two vice presidents of sales, one for retail and one for food service outlets (restaurants, hotels, fast-food chains, etc.), were hired, and a crop was readied for harvest in October. As summer 1992 waned, Calgene's outlook on its tomato, in terms of the U.S. regulatory process anyway, was very bright indeed.

By Labor Day, however, clouds started rolling in on what would turn out to be a very dark year. It started with a fire that swept through one of Calgene's three major labs over the holiday weekend. No one was hurt, but the loss of equipment and, worse, laboratory notebooks full of data was quite extensive. But, as bad as the fire was, it seemed relatively insignificant compared to the news I got from Bill Hiatt a few days later concerning our wholesomeness studies.

He had said he had a "special assignment" for me as we approached a lab bench piled high with genetically engineered tomatoes. "We're going to be running more rat feeding studies," he said, staring at a fruit he held in his hand.

"More? Why?" I asked.

He mumbled something about being on the safe side and went on to explain that he wanted me to personally prepare the fruit for shipping (a task usually carried out by research assistants) and to make sure that it was free of any pesticide residues. "Wash each tomato really well," he told me as I stood there with my mouth open. But before I could get out my "What's going on here?," he blurted out the whole story.

"I can't keep you in the dark about this," he confessed. "There's a problem with the rat feeding studies."

I was flabbergasted. "What kind of a problem?"

"The rats fed Flavr Savr tomatoes developed some kind of stomach lesions that the control rats didn't. We're going to repeat the tests."

The Troublesome
Wholesomeness Study

Bill was referring to the results of the second wholesomeness study, the one that had been initiated soon after the FDA published its statement of policy. They were unnerving. Minor-looking lesions were noted in the stomach linings of 4 of 40 rats that had been fed one particular variety of Flavr Savr tomato. None of the other rats in the study "fed" water, nontransformed tomatoes, or another variety of Flavr Savr tomato developed any lesions. And none of the 120 rats in the 28-day wholesomeness study conducted in April had developed any lesions either. There seemed to be no escaping it. These data appeared to indicate that one of the three Flavr Savr tomato varieties we had tested for unknown, unintended, unexpected effects had one: variety CR3-623 apparently caused lesions to form in the stomachs of rats.

My initial shock at the rat results didn't last long. With the plant 502-2021 incident, I had resisted jumping to the erroneous conclusion that Flavr Savr genes hop around in the plant genome from one generation to the next. In this case, too, I figured there had to be another explanation. We just had to find it. Only this time it wasn't going to be as easy as looking over an old field trial map.

But we wasted no time worrying about the degree of difficulty in the task before us. All the available fruit from the offending variety and its nontransformed parental line was harvested immediately and prepared for additional tests. Besides carefully washing the fruit to eliminate the possibility that pesticide residues might have caused the problem, these preparations included concentrating fruit samples by freeze-drying them. If the lesions were really related to this tomato, then feeding rats twice the original amount of fruit should cause a corresponding increase in the number of stomach lesions formed. An additional wholesomeness study of variety CR3-623, which would replicate the original as well as test for this dosage effect using the freeze-dried tomatoes, was initiated September 11.

Samples of fruit were also earmarked for extensive glycoalkaloid analysis. We wanted to test the hypothesis that elevated levels of these toxins in variety CR3-623 might be responsible for the stomach lesions. Unfortunately, John Uhlig, who had done the original glycoalkaloid analysis for the advisory opinion document, was no longer conveniently working in Davis at the Campbell Institute for Research and Technology. I tracked him down, but he was unavailable for subsequent analyses. I scoured the glycoalkaloid literature for other names and made phone calls, but to little avail. Experts in tomato glycoalkaloid research were nonexistent. The model system of choice for essentially everyone who analyzed glycoalkaloids was potatoes. No one, as far as I could discern, had a routine method of analyzing tomato glycoalkaloids in the fall of 1992.

Our choices were either to develop and utilize this innovative agricultural technology ourselves, true to the Calgene Fresh statement of purpose, or to farm it out to a university lab specializing in potato glycoalkaloid analysis. We farmed. I found a professor at the University of Maine willing to develop analytical methods for tomato and then use the newly forged techniques to determine the levels of these compounds in Flavr Savr tomato varieties.

Don Emlay's response to the CR3-623 wholesomeness study results was to hire pathologists from a Maryland company called PATHCO, Inc., to reevaluate the results of all three 28-day feeding studies. This Pathology Working Group (PWG) would then provide Calgene with its conclusions as to the cause of the stomach lesions.

Don also reenlisted the services of ENVIRON Corporation. ENVIRON assembled a five-member scientific panel of food safety experts. Don, Bill, Keith, and I met with these experts to review the findings from PATHCO's pathologists at a meeting held at ENVIRON on December 17.

At first glance, the PWG's results weren't encouraging. They identified lesions, which they described as erosions of minimal to mild severity, not in just the four original but in four additional rats that had been fed Flavr Savr tomato variety CR3-623 in the second study. The frequency of rats in that experimental group exhibiting erosions thereby jumped from 10 to 20 percent. The careful reexamination carried out by the PWG also revealed another three erosions not previously noted in the first two studies. As luck would have it, all three were in rats fed some variety of Flavr Savr tomato or another and none in rats fed nontransgenic tomatoes or water.

But the results of the third study, with or without reevaluation by the PWG, were cause for everyone at the meeting to relax, at least a little. Yes, erosions were present in the stomachs of a few rats fed Flavr Savr tomato variety CR3-623 in the third study also. However, erosions were also found in the stomachs of rats fed nontransformed CR3 tomatoes and of those given nothing but distilled water for the duration of the study. In fact, the largest percentage of erosions was found among the water-fed control group. What's more, there was no evidence of a dosage effect. Fewer erosions were identified in the rats that were fed concentrated tomato puree than in those fed puree from tomatoes that had not been freeze-dried.

These results, together with the fact that the University of Maine professor had not detected any tomatine, the most prevalent glycoalkaloid in tomatoes, in CR3-623 tomatoes, all suggested that the erosions were not treatment related, and that was in fact the conclusion reached by the PWG. The consensus among the experts at the ENVIRON meeting was that the lesions occurred randomly and with highly variable incidence and were therefore simply spontaneous in origin. But, we all wondered, would the FDA agree with our assessment? Were results from one "good" study enough to negate the "bad" results? Or would it take 100 "good" studies to eliminate suspicion from Flavr Savr tomato variety CR3-623? Of course, no one in that room, including those who had previously worked for the FDA, could answer these questions and nobody tried. One thing was certain. The prospect of carrying out any more wholesomeness studies was most unappealing. The group's plan of action, therefore, didn't call for any.

Bill, Don, Keith, and I were already in the midst of preparing a large package of data in response to four pages of questions the FDA had sent to Calgene earlier that month. Our colleagues at ENVIRON and the gathered panel of food safety experts agreed with Bill that Calgene should submit a thorough analysis of Flavr Savr tomato variety CR3-623 along with the data answering the agency's specific questions. That thorough analysis would include the three original 28-day rat feeding studies, the PWG's report on those studies, glycoalkaloid measurements, a literature review of glycoalkaloids and tomato fruit safety, and a comparison of the historical incidence of gastric erosions in control experiments. The food safety expert panel would then review all the information and write a summary of their conclusions about the safety of the Flavr Savr tomato for inclusion in the package prior to its submission to the FDA. I left the meeting believing that we'd found the explanation we'd been looking for and hopeful that this next data package would be the last one we would need to send to the FDA.

The FDA Responds to Calgene's Request for Advice

In its letter to Calgene of December 2, 1992, the FDA asked for additional information in order to continue its review. Some of the information the agency asked for in that letter was essentially procedural. The FDA wanted the product specification sheet for the purity of the APH(3')II protein and the reaction conditions used to perform an immunoblot experiment, for example. It also wanted copies of a couple of scientific papers that we had referred to in the Flavr Savr tomato submission but not previously provided.

Several of the FDA's other questions took a little more effort to address. The agency wanted data gathered from additional generations of the eight original Flavr Savr tomato varieties we'd analyzed. Rick Sanders and I therefore finished up our experiments demonstrating that the Flavr Savr gene was stable over five generations. We also showed that vitamin A and C levels remained normal in fourth-generation Flavr Savr tomatoes stored over extended periods.

The FDA also asked for an analysis of every putative protein coding region and every possible bacterial gene promoter that was present on the piece of *Agrobacterium* DNA that was transferred into Flavr Savr tomatoes. We had touted the precision of the technology in our requests of the agency. The agency obviously wanted precision of us in return. So, Ray Sheehy and I prepared a DNA base pair by DNA base pair analysis of the transferred DNA, the T-DNA, and explained why none of these DNA sequences, other than the ones comprising the Flavr Savr gene and *kan*^r, was expected to be expressed in a Flavr Savr tomato or any other organism.

I felt all of these questions reflected the fact that the agency was conducting a thoughtful, methodical, and scientific analysis of the data Calgene had submitted. The FDA's scientists were asking the same questions I would have asked of Calgene had I been in their lab coats, but for one. The first question raised by the agency in their December letter was "whether only the DNA sequences within the T-DNA region were integrated into the tomato genome." In other words, were any *Agrobacterium tumefaciens* bacterial genes or other vector sequences outside the so-called T-DNA region included along with the antisense PG gene in the segment of DNA that was transferred into the recipient plant's genome? I would never have thought to ask that question.

The fact that only T-DNA from *A. tumefaciens* was transferred into a recipient plant's DNA was supported by 10 years of scientific studies in the field of plant molecular biology. There was no question in my mind about it; transfer of any DNA other than T-DNA just didn't happen. Even entertaining the idea seemed preposterous to me. So, instead of checking experimentally to verify that only T-DNA had been transferred into Flavr Savr tomatoes, I looked to the most reputable recent review article on *Agrobacterium* I could find. In it, Patricia C. Zambryski, an expert on *Agrobacterium* from U.C. Berkeley, explained T-DNA this way: "The T-DNA is defined and delineated by . . . the T-DNA borders. Any DNA, and only DNA, *between* [emphasis added] these borders is transferred to the plant cell."[29] The two-paragraph long answer I gave to the question at the top of the FDA's list, including this quote from Professor Zambryski, was simply a literature review.

Keith Redenbaugh assembled the answers to this and the FDA's other questions together with the thorough safety assessment of Flavr Savr tomato variety CR3-623 as had been proposed at the December meeting at ENVIRON. The several hundred pages of information were filed with the FDA on March 1, 1993. The expert panel's conclusion, included in the assembled document, was "that Flavr Savr tomatoes are as safe for human consumption as other tomatoes that are currently part of the human diet." Calgene made sure that the panel's conclusion, as well as the rest of the rat story, was also reported in the newspapers.[30] The company projected (and I continued to hope) that the stack of documents filed March 1 "should answer the last of the FDA's remaining questions on the safety of the Flavr Savr."[31]

The Steep "Learning" Market Curve

During the 6 months I'd been distracted with wholesomeness studies, the slope of the learning curve associated with Calgene Fresh's vine-ripe tomato test market became apparent. To use skiing parlance, it was a double black diamond. The idea originally had been that, once Calgene Fresh started selling MacGregor's premium tomatoes, their quality and quantity would never be allowed to falter. But by late summer 1992, when one might expect to find lots of tomatoes of high quality, Calgene Fresh had an inferior product on supermarket shelves. The quality of MacGregor's tomatoes got so bad that in-house they were being called the "cream of the crap."

Shrink, the number of unusable tomatoes, was also running very high. As much as 50 percent of the available fruit was being thrown away. Tomatoes were discarded in the field, on the packing line, and in grocery stores because they were not up to MacGregor's standards, not even "cream of the crap" standards.

In-store shrink resulted from a sales incentive program Calgene Fresh had implemented. In order to obtain retail space for MacGregor's tomatoes, the company agreed to buy back any fruit that a grocery store did not sell. Produce managers took full advantage of the program, often returning over 20 percent, and sometimes as much as 60 percent, of the fruit delivered throughout most of fall 1992. Similarly, only 10 percent of the first delivery of MacGregor's tomatoes to a food service outlet that September was deemed acceptable by the customer.

Because of these high levels of shrink, costs were out of control. In late August 1992, the retail price for MacGregor's tomatoes was running about $1.39 a pound, while the cost of delivering those tomatoes to market was approximately $5 a pound. Keeping the promise of providing tomatoes with consistently superior texture and flavor was turning out to be harder and vastly more expensive than anyone at Calgene Fresh had originally thought.

But the company was still conducting its "learning" market and, relatively speaking, it had learned quite a bit during the first 5 months it had been selling tomatoes. Back in April, for example, the cost per pound to deliver MacGregor's tomatoes had been $9.50, nearly twice as much.[32] And reducing the cost of goods by half had been accomplished without an official partnership with an experienced tomato grower-shipper or any Flavr Savr tomato–related handling advantages. By August 1992, however, it was time to take full advantage of both.

That's when arrangements with three tomato supplier-shippers, two in Florida and one in California, were finally formalized. The basic arrangement called for the grower-shippers to produce Flavr Savr tomatoes and ship their crops to Calgene Fresh, where marketing and sales would be handled.[33] The tomato-growing locations of Taylor & Fulton, Inc., Gulfstream Tomato Growers, Ltd., and Meyer Tomatoes, Inc.—especially since Meyer, one of the largest tomato shippers in the western hemisphere, also had growing operations in Mexico—could provide Calgene Fresh with a year-round supply of fruit. And each producer had indicated a willingness to modify their harvesting and packing operations in order to achieve MacGregor's-level quality.

The deals were a tremendous coup for Calgene Fresh. These companies provided expertise where Calgene Fresh needed it most, and working with them gave Calgene Fresh its only real chance at fulfilling the promise of providing high-quality tomatoes year-round. And the primary thing these partners wanted in return, as described by Jay Taylor, president of Taylor & Fulton, was to be part of a "new era in produce . . . a phenomenal deal . . . a legitimate step forward for the consumer."[34]

The establishment of these tomato-growing partnerships marked a turning point in the "learning" market process for Calgene Fresh. With the help of its partners, Calgene Fresh could now test, on a production-level scale, the 4-year-old hypothesis that vine-ripened Flavr Savr tomatoes could be produced and handled like mature green fruit. And although the company continued to test the market with non–genetically engineered tomatoes, especially with regard to pricing, it also began to focus in earnest on the specific production and handling needs of Flavr Savr tomatoes. A team of Meyer Tomatoes and Calgene Fresh employees was immediately assembled to design a series of experiments to demonstrate the most effective way to manage shrink reduction.

A Series of Business Snafus

On December 10, DNA Plant Technology Corporation (DNAP), clearly one of Calgene Fresh's most significant competitors, announced that it had entered into an agreement of its own with Meyer Tomatoes, the largest of Calgene Fresh's shipper-grower partners.[35] Meyer would cultivate, harvest, package, and ship DNAP's superior-tasting, vine-ripened, proprietary VineSweet tomato. And, DNAP emphasized in its press release, the VineSweet tomato was developed utilizing a phenomenon called somaclonal variation, in which plants regenerated from individual cells to which specific genes have *not* been added via "transformation" are screened for characteristics exhibited simply as a consequence of undergoing the tissue culture process. Therefore, it would not require U.S. regulatory approval. The company, with the help of Meyer Tomatoes, planned to carry out its initial market entry with VineSweet tomatoes sometime during the first half of 1993.

Although Meyer Tomatoes and DNAP had already been producing and comarketing DNAP's patented VegiSweet minipeppers through a joint venture called Freshworld Farms® and Calgene Fresh management knew DNAP had VineSweet tomatoes in its commercial pipeline, the VineSweet announcement still came as a complete surprise. All the adaptations of methods and equipment that Dan Wagster, Matt Kramer, and colleagues had been working on with Meyer over the previous 4 months would be just as important for the handling of VineSweet tomatoes as for Flavr Savr tomatoes. And, as if to twist the knife, Bob Meyer was quoted in DNAP's press release as saying, "The VineSweet tomato, to us, is the best-tasting tomato we have tried."[36] As when the Campbell Soup Company, after funding much of the Flavr Savr tomato's early development, announced that it would not use genetically engineered tomatoes in any of its products, at least not until consumer acceptance of biotech foods had been established,[37] Calgene employees felt betrayed. The DNAP–Meyer Tomatoes incident left Calgene Fresh reeling.

Tom Churchwell and Dan Wagster met with Bob Meyer a few days later to clarify the partnership and determine its future. But they brought very little ammunition with them. MacGregor's tomatoes, while somewhat better than they had been in August, still lagged behind the competition in terms of quality. Jim Gerecke, Calgene Fresh's VP of retail sales, had recently described their flavor as worse than that of other grocery tomatoes. Calgene Fresh simply needed Meyer Tomatoes more than Meyer Tomatoes needed Calgene Fresh. For one thing, the first big shipping test of Flavr Savr tomatoes out of Mexico was only a few weeks away. Therefore, the partnership, such as it was, continued despite the DNAP deal. Bob Meyer continued to cover his bets.

The shipping test out of Mexico, however, proved to be yet another disaster. It was designed to test, not only whether the Flavr Savr gene would enable vine-ripened fruit to survive 2000 miles in a truck, but also whether Flavr Savr tomatoes could be bulk-loaded into large bins, like the ones often used to display watermelons, and sorted and packed at destination, not before. A group of Calgene Fresh employees, including Matt Kramer, Dan Wagster, Steve Benoit, and Ken Moonie (brought on board at Calgene Fresh the previous September as director of finance and business development) was anxiously waiting when the truck arrived in Chicago.

The results of the test were clear before the vehicle had come to a complete stop. Tomato puree seeped from the truck's back end. The watermelon bins had collapsed upon each other. The cargo was completely unsalvageable. As Steve muttered repeatedly, "It's over, it's all over," Matt and Ken used snow shovels to transfer the mess to dumpsters. Dan Wagster, looking decidedly paler than before the truck's arrival, stood by in his three-piece suit reflecting on the half-full glass of water.

"We're learning," Dan said. "It's just part of all the learning we're doing."

"All we're learning," Ken replied, "is how to shovel god-damned tomatoes." And that, unfortunately, was just the beginning. Tomato-shoveling and dumpster-filling skills would continue to be honed at Calgene Fresh over the next several years.

Then, a few days before New Year's, Jim Maryanski called to say that the FDA preferred to regulate *kan*^{r}, or specifically APH(3')II, as a food additive after all. Roger Salquist, Don Emlay, and I took the call on the speaker phone in Roger's office. Government officials, including members of Congress, were more familiar with the food additive petition (FAP) than the advisory opinion process, and therefore, the reasoning went, an FAP route would likely be smoother. Of course, Environmental Defense, the National Wildlife Federation, and Jeremy Rifkin had also all advised the FDA that the appropriate way to handle *kan*^{r} was via an FAP, but no one mentioned any of those groups during that phone call.

Roger Salquist reacted to the news with understandable concern. Two years had passed since the original document had been filed with the agency. Time had been of the essence then; it was even more crucial now. In addition, Roger was worried about the "road show" he was embarking on in January to raise more capital. How would this news affect Calgene's stock price?

Jim Maryanski reassured Roger that, since Calgene's advisory opinion request was in the FAP format anyway, refiling the information was simply a formality. Calgene would lose no time. The FDA's ongoing review of the company's data would be unaffected. The kan^r document was therefore officially resubmitted as an FAP on January 4, 1993.

What was of concern for the stock price, it turned out, was news Roger received soon thereafter. In addition to the broad antisense patent it had issued to Calgene, the U.S. Patent Office had issued several other very broad antisense patents to the State University of New York that had subsequently been exclusively licensed to Enzo Biochem, Inc. Suits and countersuits alleging patent invalidity, patent infringement, patent noninfringement, and unfair competition, among other issues, were then initiated by both Calgene and Enzo.

In response to Enzo's suit alleging willful patent infringement by Calgene, all Calgene and Calgene Fresh employees, some 150 at the time, were instructed to turn over any files, papers, or notebooks that referred in any way to the use of antisense technology. It was a big job. The company had over 2300 lab notebooks at the time, for example, and half of them or more likely contained antisense-related work.

Since it was already common knowledge that Calgene was using antisense technology, the assignment also seemed an incredible waste of time and trees. Shouldn't the effort be concentrated on establishing the validity of each patent rather than the infringement issues, I wondered? Nevertheless, we all went through our offices and lab benches with fine-tooth combs and turned over our reams of paper, truckloads altogether, to the attorneys.

But, in spite of the product development dilemmas, rat regulatory issues and patent problems—difficulties in all of the areas that had originally made the Flavr Savr tomato seem as good a business opportunity as NutraSweet had been—Tom Churchwell remained unabashedly bullish on Calgene Fresh. That fact was made very clear to me during an executive staff meeting in the Evanston office that January. The meeting was called to discuss the draft of a 5-year strategic plan, and Tom, in the interest of open communication, had invited me, the only nonexecutive in the room, to listen in. He opened by asking if anyone had any questions or comments on the plan.

I was aware that Tom and his vice presidents had been working with a consultant from Harvard Business School, Ben Shapiro, while they put together their plan because, also in the interest of open communication, information about nearly everything going on at Calgene Fresh was sent or e-mailed to nearly everyone else in the subsidiary. I looked down at the 5-year plan in front of me. I looked at the vice presidents sitting quietly around the table and especially at Bill Hiatt next to me. No one said a word. I raised my hand.

"This plan calls for a large-scale national rollout of tomatoes. I thought Ben Shapiro recommended filling a small niche market first and expanding later."

"Ben Shapiro is a really sharp guy," Tom responded with a smile, "and I agree with everything he says . . . except on this issue." End of discussion.

I'd gotten off easy, perhaps because I was a scientist and Tom felt I didn't know any better. Ken Moonie had received a considerably more negative response from Tom just a few weeks earlier when he suggested that Calgene Fresh would be a $50 to $75 million business "tops." The trouble was that, although Steve Benoit's original business plan had called for the company to be making $150 million in sales within 7 years, Tom had suggested to his management team that Calgene Fresh could be a billion-dollar company. The visionaries on the team bought into the idea, and it soon became only too obvious that they didn't want to hear anyone say otherwise. The realists became part of the problem, not part of the solution.

On the plane ride home, I asked Bill why he hadn't questioned Tom about the scale of the rollout. "What good would it do?" he replied. I didn't take Bill's hint. I continued to naïvely adhere to Calgene Fresh's "Core Values and Beliefs," which called for "voicing of opinions" and "question[ing] company decisions." Nevertheless, I did try to bury myself in regulatory affairs as much as possible for the next couple of months.

Skunk Works

Tom Churchwell's search for quality brought a group of management consultants espousing the virtues of "skunk works" to Calgene Fresh in the spring of 1993. Skunk works refers to Lockheed Martin's famous aircraft design and production facility, where talented individuals are given complete creative freedom to come up with innovative solutions for the company's projects. Tom loved the concept, especially the idea of employee-organized interdisciplinary teams. Others of us were more skeptical.

"Everyone in this company needs to learn how to do everyone else's job," Tom proclaimed at an introductory skunk works meeting.

"Shouldn't we learn to do our own jobs first?" was Ken Moonie's immediate response.

But Tom would not be swayed. The more enthusiastic Calgene Fresh employees organized skunk works meetings immediately.

Kanti Rawal, Calgene Fresh's tomato breeder, seemed to be among the most ardent supporters of the Calgene Fresh vision. He was already in the midst of developing what he called Velcro® tomato varieties in order to hit the marketplace with Flavr Savr tomatoes in the shortest possible time frame. His plan called for bypassing traditional breeding steps and "throwing tomato varieties together to see what sticks." Kanti organized one of the first Calgene Fresh skunk works, a day-long workshop that he called "MacGregor Quality: Doing It Right the First Time and Forever."

Half a dozen experts from research labs around the country gathered in Davis on March 3, 1993, and gave talks on subjects ranging from the biochemistry of tomato flavor to how humans integrate the five senses during taste tests. As a scientist, I found the research all very interesting. But it was hard to imagine how that vast amount of basic information was going to help Calgene Fresh deliver quality tomatoes in the short term.

Dan Wagster saw it differently. He was excited. He had been pushing to get Bill Hiatt and his R&D staff to come up with a quantifiable system for defining tomato quality ever since MacGregor's tomato quality had bottomed out the previous summer. I had agreed with him about it then; it was something our quality-driven company should strive for. Our time frames, however, were simply not on the same scale.

Dan had finally convinced Bill, I discovered during my annual performance review later that week, to develop quantitative "product specs." I was to head up the effort in conjunction with evaluating the company's second potential product, a fresh tomato in which the expression of a host of ripening genes could be attenuated by "antisensing" production of the plant hormone ethylene. I definitely had my work cut out for me, but the basic nature of the research was a welcome change.

On March 17, I held my own skunk works, a product specs meeting in Evanston. I presented a plan to test a large number of tomato varieties for correlations between taste preferences and various quantifiable characteristics, such as volatile compounds. If certain compounds or combinations of them were always associated with the tomato varieties people found most flavorful, we would be in business. Theoretically, once established as a standard, a profile of flavor-associated compounds could then be used to select tomato varieties for product development as well as demonstrate the superior flavor of our tomatoes over the competition's.

Dan Wagster loved the idea. It would represent a vast modernization over the methods used at the time to select flavorful tomato varieties, the most successful being Jay Scott's breeding program at the University of Florida. Scott had described at the earlier quality workshop how he and his assistant tasted tomatoes in the fields. One of the two liked more sugar in their tomatoes, the other more acid. If neither or only one of the two liked a particular variety, it was thrown out of the program. If both liked it, it remained. Solarset, Scott's successful variety, had been selected in this manner.

Dan wanted something much more concrete than Scott's system, though. He wanted numbers. A proprietary profile of volatile compounds correlated with acids, sugars, and other components of tomato flavor would fill the bill nicely, and he had absolutely no problem with the test's $200,000 price tag. But, although he was not particularly pleased when I told him it would take 6 months to complete the test, he really became unhappy when I explained that it was possible we wouldn't identify any correlations. It was an experiment, after all.

Linda Gohlke, an attorney and acquaintance of Tom's from his NutraSweet days, asked me at lunch a few days later, "Do you think that, when you say you can't predict the outcome of an experiment, Dan might hear that you don't care about the experiment's results?" I was too dumbstruck to answer. I knew I still had a lot to learn about business, but could it be possible that Dan, having spent 6 years at Calgene by then, still didn't understand the fundamental definition of an experiment?

In contrast to Dan's negative response to my experiment, members of the Calgene Fresh business staff could also react to research results with undue enthusiasm. One such occasion involved an experiment on tomato storage conditions conducted by Mike Boersig, a Ph.D.-level scientist who, like Keith Redenbaugh, had come to Calgene via the merger with Plant Genetics, Inc. Mike found that tomatoes stored in high humidity weighed more after storage than before. Members of the business staff were ecstatic. Heavier tomatoes, which were sold by the pound, meant bigger sales. In multiple e-mails sent to everyone at Calgene Fresh, Mike was hailed as a "genius."

But the condition of those heavier tomatoes got lost in all the ecstasy. Their increased mass was due not only to water weight. Increased numbers of bacterial and fungal organisms had also flourished on them in the humid conditions. Rather than the increase in "$" referred to in the congratulatory e-mails, the humid conditions used in Mike's experiment were more likely to increase tomato "shrink."

The Calgene Fresh science staff was also asked to remove the error bars from graphs comparing MacGregor's tomatoes with the competition. The business staff found them too "confusing." Coincidentally, MacGregor's tomatoes often appeared superior to the competition's when the error bars, which indicated that there was no statistical difference between them, were eliminated. Despite our early interfacing meetings, the business and science cultures at Calgene Fresh were obviously still worlds apart.

Agrobacterium: An Act of God

I couldn't dwell on the difficulties presented by the cultural abyss at Calgene Fresh for long. In a letter dated April 6, the FDA let us know that my literature review answer to its question about T-DNA was insufficient. The agency wanted Calgene to specifically supply any experimental evidence it had that demonstrated that only the T-DNA had been integrated into the tomato genome.

I had misinterpreted the FDA's request. We had lost precious time because of it. I tried to apologize to Bill Hiatt over the incident, but he wouldn't accept my apology. In fact, he placed the blame with himself. He had been the one originally asked the question on the phone, he told me. Therefore, he should have realized that the FDA wanted experimental evidence. We had had a couple of run-ins, primarily over staffing issues, during the previous months that had left me wondering whether Dr. Hiatt and I were still operating on the same wavelength. I left his office more enthusiastic and confident than when I entered, feeling reassured that we were.

Bill's shot of enthusiasm lasted long enough to see me through the execution of our first experiment designed to demonstrate that only T-DNA was inserted into Flavr Savr tomatoes. Unfortunately, there was nothing Bill could do or say to deflect the despondency I felt when he told me the results of that experiment.

We were traveling between Calgene Fresh, Davis, and Calgene Fresh, Evanston. We hadn't talked on the plane because Bill, who made many more of these trips than I did, had used frequent flyer miles to upgrade to first class. As we were negotiating the subterranean moving walkway system that connects the two United Airlines terminals at O'Hare, Bill asked for my reaction to the T-DNA experimental results.

"I haven't developed the Southern blot yet," I replied.

"Rick developed it this morning," he said matter-of-factly, "*Agrobacterium* vector sequences were present in several Flavr Savr tomato lines."

I stopped in my tracks (although I kept moving). It couldn't be true. The flashing lights and recurrent United theme song intensified the surrealism I felt.

Bill kept right on walking. "We'll meet with the regulatory task force when we get back to Davis," he said over his shoulder. "We'll figure out what to do about it then."

The regulatory task force meeting was held in Calgene's boardroom. To say that those in attendance were tense would be putting it mildly. Of great concern was the fact that the data sent to the USDA nearly a year before— which the agency had used to determine that the Flavr Savr tomato was not a plant pest and would no longer be regulated as one—contained false information. At the time the data were submitted, we had all believed, of course, that only T-DNA had been inserted into our tomato varieties. But we now knew that some 20 to 30 percent of our tomato varieties contained additional DNA, including an additional antibiotic-resistance gene, that the USDA would certainly have wanted to consider before making its final determination. In comments the USDA originally made on the scope of its interpretive ruling on Calgene's petition, the agency specifically stated that "authority could be reasserted if a new plant pest risk should ever be uncovered in the future." Reasserting its authority was about the last thing anyone in that room wanted the USDA to do.

Some task force members were worried the whole situation would look like a cover-up. Don Emlay declared that he had stopped writing anything down. He would no longer take notes. He wouldn't even write in his personal diary anymore. David Stalker described *Agrobacterium* as "an act of God."

After the venting subsided, we settled on a plan. Rick Sanders and I would analyze the more than 70 Flavr Savr tomato varieties in product development for the presence of bacterial DNA from outside the T-DNA borders. Those plants that contained any would be eliminated from the commercialization pipeline. In the meantime, all genetically engineered tomatoes would be grown under the contained conditions the USDA required of regulated articles, and we would provide our experimental evidence to both the FDA and the USDA ASAP.

I left the room wondering whether this "act of God" was really ready for the marketplace. The T-DNA revelation had shaken my faith in the technology.

The Beginning of the End of Flavr Savr Tomato FDA Science

Later that month, the entire Calgene Fresh R&D team held the first of a half-dozen or so all-day, off-site skunk works meetings. We had put together our own strategic plan a few weeks earlier, and the specifics of prioritizing projects and deciding just who would do what was to be worked out at these gatherings. I wasn't looking forward to them. I much preferred to discuss the scientific plans and results themselves rather than the organization around them. Besides, I had a lot of plants to analyze for evidence of acts of God.

The group decided that regulatory science was still the top priority for Calgene Fresh R&D. I didn't disagree. But I realized during the course of the meeting that I no longer wanted to head up the regulatory science effort. I had rationalized my reduced faith in ag biotechnology as burnout by then. I simply couldn't evaluate the antisense ethylene plants, oversee the tomato flavor-quality experiments, and keep up the regulatory effort, especially if Bill continued to deny me additional staff. Something had to give. I wanted it to be my participation in regulatory science.

I spoke to Bill about my situation the following Monday morning. He seemed sympathetic. But, since we were making these decisions by group consensus, he left it up to the group to decide.

By the end of the next R&D meeting, the group's priorities had all been established and many of the staff assignments made. I had drawn considerable attention to myself by my silence during a minimal discussion of the regulatory effort. Regulatory staffing remained unassigned. As we gathered our things to go, Bill announced that "whoever comes late to the next meeting will have to do regulatory." In one fell swoop, the group's number one priority became the booby prize.

I couldn't read Bill at all anymore. Was he simply mocking (and thereby negating) the process we had collectively just spent 400 person-hours on? Or was he deliberately discouraging my colleagues from taking on the regulatory science load? Whatever the reason, I had reached my limit. It was bad enough that MacGregor's tomato quality was still only a dream (MacGregor's tomato flavor had been recently described in-house as "grassy, sometimes musty") and that Dan might be having difficulty differentiating an experiment from an expenditure and that I had serious qualms about how Tom was running the business generally. But if I didn't have the support of my immediate boss in the chaos around me, I needed to find another job. I cogitated over the situation throughout the weekend and discussed my career options with a professor from U.C. Davis at lunch on Monday. I gave Bill my letter of resignation that afternoon.

As with my apology over the T-DNA situation, he refused to accept it. "Explain the issues to me again," he asked. I went over the burned-out, understaffed scenario I had discussed with him just 2 weeks earlier. He surprised me by asking for a list of what it would take to keep me. He seemed willing to support me after all.

Bill also had me talk to Tom Churchwell about it. I met with Tom in Roger Salquist's office a couple of days later. The discussion was typically Churchwellian. He talked in what sounded like circles, and I asked for clarification occasionally. At one point he said that when he was talking to people he was usually selling. "But not now," he told me, "not here talking to you." He seemed to be trying. I believe his intentions were good. My decision to stay at Calgene Fresh, however, was made in spite of Tom Churchwell.

My list for Bill was short. I wanted out of regulatory science. I wanted to clarify a specific supervision situation. I wanted to cut my workweek to 80 percent in order to spend more time with my 14-month-old daughter. Bill, with a smidgen of hesitation about the cut to 80 percent time, accepted my terms.

I'd already finished preparing the experimental data on the T-DNA issue. We sent it to the FDA on June 4. Keith had put together a revised environmental assessment for the *kan*[r] FAP, as requested by the agency, and sent it in the day before. Except for exploring the possibility that APH(3')II in canola and cotton seed products might inactivate antibiotics added to animal feed, an issue being handled by Calgene, Inc., employees working with those crop plants, everyone at Calgene Fresh hoped we had now answered all of the FDA's questions about the Flavr Savr tomato. With a little luck, the next time anyone at Calgene Fresh did any regulatory science it would be in preparation for launching the company's next genetically engineered product.

The Calgene Fresh R&D group agreed that I should head up the antisense ethylene product development program. Bill did, too, saying "You run ethylene, and you make the calls." I looked forward to a scientifically productive summer.

The End of the End of Flavr Savr Tomato FDA Science

My respite from regulatory involvement ended just a few days later. Don Emlay received another letter from Linda Kahl at the FDA on June 8, this one indicating that the agency was unhappy with the glycoalkaloid measurements included in our March 1 submission. Because the measurements were made using unpublished methods, the FDA wanted a thorough, publication-quality description of the analytical methodology that had been used.

Don, Bill, Rick Sanders, and I met to discuss the situation that afternoon. It went without saying that I was back on board the regulatory team. Bill's positive response to my career crisis had somehow rekindled my enthusiasm. I was on the phone to the University of Maine that afternoon.

Several weeks later, gasoline was thrown on the regulatory fire we were trying to put out, when Don received yet another letter from Linda Kahl. The agency was concerned about the differences in the number of stomach lesions observed in the three wholesomeness studies. The FDA wanted additional data or information that could provide an explanation for the disparity among the three studies and scheduled a meeting with Calgene in early July to discuss the agency's overall review of the animal studies.

At the July meeting, the FDA indicated that, despite the fact that we were no longer going to commercialize the troublesome Flavr Savr tomato variety, CR3-623, the agency wanted a thorough, repeatable analysis of the glycoalkaloid compounds in that line. Matt Kramer, who had gotten early word on the CR3-623 decision, called from the O'Hare airport to forewarn me that "Bill will be coming at you with both barrels."

There was no denying that Bill was in crisis mode the next day, but he viewed the crisis as scientific, and Bill was at his best in a scientific crisis. Don Emlay tried to interject that science wasn't the only answer; the approach to take was also part "an art, not a science." Bill vehemently disagreed. The interaction marked a low point in the relationship between the two men. They rarely spoke to one another for months afterward. (Keith Redenbaugh admirably served as the go-between for regulatory relations between Calgene, Inc., and Calgene Fresh during that critical time.)

Bill's plan called for a scientific response, and he wanted the science in that response to be "superb." He was shooting for September 1 to submit all the data. That gave us about 6 weeks.

The rest of the summer was a blur. Every day I was either calling our glycoalkaloid expert in Maine or sending fruit to Maine or in Maine myself trying to ensure that the glycoalkaloid analysis went smoothly. We even sent a Calgene Fresh research assistant, Steve Vanderpan, to the University of Maine in order to speed up the analysis.

The final raw glycoalkaloid data from the University of Maine arrived at my home the Saturday before Bill's deadline. I spent the rest of the weekend deciphering it but had concluded by Monday morning that, if Bill still wanted to submit the data on September 1, we didn't have enough time to get it, in its entirety, into "superb" form. As an alternative, I suggested we summarize the results and include the raw report as an appendix. Bill disagreed and was visibly angry with me over it. He assigned Matt Kramer to write a complete and clean report. Matt gave me his draft late on Tuesday. I came in the next morning on what was normally my day off to go over it with him, but it was still nowhere near ready to go on that, Bill's deadline, day.

By the end of the next day, the report addressing the disparities among the rat feeding studies, which emphasized that the data did not support a test material–related cause for the stomach lesions,[*] had not been finalized either. Bill pleaded with Keith to submit it before midnight. Keith obliged. He finished putting it together by 10 PM and then drove to the Sacramento airport so the document would make the last Federal Express flight out that night. When Keith finally got home, he received a phone call informing him that his dad had died after the surgery he'd undergone earlier that evening.

---

[*] A conclusion corroborated by the FDA's Center for Food Safety and Applied Nutrition during its final assessment of the Flavr Savr tomato and by the European Union's Scientific Committee on Food after its review of Calgene's data (as reported in October 2000).

Bill didn't come in to work the next day. Matt and I reluctantly called him at home about the glycoalkaloid response for the FDA. To say he was upset would be putting it mildly. "Just send it in," he told Matt. We sent it in.

Two weeks later we received the inevitable call from the FDA. The agency still wasn't satisfied and wanted still more details regarding the glycoalkaloid analysis. This time we brought our glycoalkaloid researcher from Maine out to California. He was available for consultation as we put the final requested data on Flavr Savr tomatoes together for the FDA. We filed the information with the agency on October 1, 1993.

The FDA let Don know soon thereafter that the agency had no further questions about kan^r or the Flavr Savr tomato. It felt as if a noose had been removed from my neck. The week after we submitted the last data, I spent a whole day checking out a field trial of antisense ethylene tomatoes and a facility designed to test them. I spent another whole day working on a manuscript at home. By the end of the week, however, the adrenaline was wearing off, and, as often happened after my final exams in college, I got sick. What was a slight upset on Thursday turned into full-blown giardiasis over the weekend. Unfortunately, I felt well enough on Monday to come in to work. Had I known what was in store for me there, I would definitely have stayed at home.

The Inquisition

I had two meetings scheduled for that morning. An R&D management team–building meeting at 10 and the usual Calgene Fresh Monday morning staff meeting via conference call at 9. As I hurried around the building trying to locate the roving staff meeting, I stumbled into what looked like the team-building gathering.

"What's going on?" I asked, "I didn't think this meeting was until 10."

No one said anything. No one even looked at me. As the moment turned awkward, I realized this meeting had all the earmarks of the personnel "interventions" I heard the visionaries on the business side of Calgene Fresh were using as a way of dealing with the realists. Finally, Pat Wheeler, the "training" consultant said, "since you're here you might as well take a seat."

Despite the several weeks we had recently spent defining our "rules of professional interaction," which included being "open and honest" with one another, the meeting had been planned without my knowledge. And the plan, it soon became apparent, called for a seven-on-one confrontation. As Pat facilitated, my closest Calgene Fresh colleagues proceeded to tell me what they didn't like about me. The recurrent theme seemed to be that I was a devil's advocate.

"But in our sessions with Pat we've discussed how a well-balanced team needs a devil's advocate just as much as it needs a compromiser and a facilitator and the rest of it," I said in defense of the role I'd been playing.

"But you play devil's advocate too much," came the response.

Mike Boersig was present but said nothing. When I asked why, he said he was there in an attempt to even the sides, if only slightly. Keith Redenbaugh and Kanti Rawal had refused to participate at all.

A couple of those who had participated thanked me for staying and interacting thoughtfully. One commented on how powerful she had felt as part of the group. After Pat Wheeler and I concluded a short debate on the definition of *inquisition*, it ended.

Bill Hiatt was out of town. Calgene Fresh's tomato partners, Meyer Tomatoes and the two East Coast tomato grower-shippers, had all severed their ties with Calgene during the previous few weeks.[38] (Coincidentally, Bob Meyer expanded his relationship with DNA Plant Technology, still Calgene Fresh's biggest rival.) So Calgene Fresh's management team, including Bill, had been preoccupied with establishing the independent contract growing, picking, packing, and state-of-the-art distribution systems to replace them.

When Bill returned, I asked him if he had known about the inquisition meeting beforehand. He answered in the affirmative. Did he have anything he wanted to say to me about it now? No. I resigned on the spot. This time, he accepted my resignation.

Don Emlay, also obviously relieved that the FDA's questions about the Flavr Savr tomato had finally all been answered, stopped by my office later that day just to ask how things were going. I told him my story. The next thing I knew, Roger Salquist came in and asked me to sit tight until he had had a chance to review the situation. I realized then how effectively Calgene Fresh had been split off from Calgene, Inc. The thought of returning to the mother ship for help had never occurred to me. I met with Roger in his office the following Monday morning.

"I've spoken with your colleagues, and they tell me that you are a contrarian," he said. He mentioned an ongoing drama in which I alone was against hiring another company to complete analyses I felt we could handle in house. Obviously, Roger had done his homework.

"Well," I said, the thought crossing my mind that a pure contrarian would disagree with him on the point, "I call it a devil's advocate, Roger, but, yes, I admit I am."

He looked me straight in the eyes. "We need more contrarians around here."

"I don't suppose you could ever work with those people again," he said, "I sure [expletive deleted] couldn't. And I can't promise you that things at Calgene Fresh will change soon. These things take time. But they will change. In the meantime, pick any other project in the company you'd like to work on and it's yours."

The End of Calgene Fresh, Phase I

At an executive retreat held in Santa Cruz, California, 6 months earlier, Roger had been 100 percent behind Tom Churchwell and his business plan despite the fact that several Calgene, Inc., vice presidents had had serious qualms about the soundness of the plan's numbers. "No more questions!" he had shouted, banging his fist on the table as Dan Wagster was being interrogated by Andy Baum and John Callahan. "We're doing this [tomato business]." As a result, much of the $27.4 million Roger had raised through the public offering of common stock in February 1993 had been made available to Calgene Fresh.

But in late October, Roger revisited the hands-off approach he had taken with the subsidiary. First, he eradicated the Calgene Fresh practice of interventions, or inquisitions. The day he delivered the message that "we will not have management by committee around here," it could be heard reverberating off the walls in the hallways of Calgene Fresh. Next, because the company was well on its way to a net total loss for the year of $42.8 million, Roger eliminated the excessive numbers of leased cars, cell phones, and laptop computers (many of which had been placed in the hands of a sales staff largely made up of kids just out of school) and the luxury of management trainers at Calgene Fresh. He considerably shortened the long list of Calgene Fresh's other paid consultants as well. Although the kind of money being spent on these items might not have seemed out of line at a company like Monsanto, Calgene didn't have a profit-generating product like RoundUp® to fall back on. The money hemorrhaging had to stop. Roger cleaned house.

The final bloodletting occurred in February 1994. That's when Roderick N. Stacey, a Calgene director since 1990 who had been serving as the company's president and chief operating officer since December 1992, was elected the chief executive officer of Calgene Fresh. Soon thereafter, Dan Wagster, Steve Benoit, and about 100 employees at the Chicago distribution facility were let go.[39] Tom Churchwell stayed on in a demoted capacity for a couple of months.

I ran into Bill Hiatt at the copy machine the morning the big management changes were announced. He was grinning from ear to ear. "The only thing that stays the same around here is that everything is always changing," he said, as he gathered up his papers and disappeared down the hall.

CHAPTER SIX

Overpromised, Underdelivered

Don Emlay took a vacation while awaiting final word from the FDA on the status of the Flavr Savr tomato. He expected a decision any day, however, so he checked in with the agency regularly throughout his trip to Boston in February 1994. During one such call, made after hours from the bar in which the television show "Cheers" was filmed, Jim Maryanski, Laura Tarantino, and Allen Rulis from the FDA informed Don that David Kessler, FDA commissioner, had decided to hold a public Food Advisory Committee meeting prior to rendering an official decision on Calgene's tomato. Don, caught unprepared on a pay phone located between the ladies' restroom and the kitchen, scribbled notes about the meeting particulars on a series of Cheers napkins.

The meeting was originally scheduled for March 8 through 10 but was postponed in order to accommodate the lead time legislated for such public meetings. On March 14, Calgene announced "that the Food and Drug Administration (FDA) has rescheduled the previously postponed meeting of the Food Advisory Committee to April 6, 7, and 8." The press release went on to say that "the committee," made up of food experts from various academic and/or scientific fields, "provides advice on emerging food safety, food science, and nutrition issues that FDA considers of primary importance in the next decade." Calgene's preparations for the meeting were already feverishly under way when the announcement was released to the press.

On March 30, Don Emlay held a practice run-through of Calgene's presentations for the Food Advisory Committee meeting in the company seminar room. He opened the rehearsal by stating that an FDA document advising that "Flavr Savr tomatoes are as safe as other tomatoes" and the "use of APH(3')II is safe" had been made public earlier that day.[1] Don therefore viewed the meeting primarily as a PR exercise for the FDA but told us that, within that context, "our product is on the line."

"We have to be absolutely accurate scientifically while saying things in a manner that can be picked up positively and understood by the press," he said. He also warned us that critics of agricultural biotechnology would be present at the public meeting and they would be well prepared.

"All the weirdos will be there," Don said. "That's their job. It's our game to lose now."

Calgene employees were quick to pick up on Don's metaphor, which he had already been espousing prior to the rehearsal. Talk in the labs centered on whether we were winners or losers. Would we end up like the Boston Red Sox in the 1986 World Series?

Consequently, the atmosphere at Don's practice session was tense. People scurried for more chairs. Don couldn't find his glasses. The slide projector wasn't working properly.

One vice president, John Callahan, tried to cut the tension by making shadow pictures with his hands in the blank white light of the malfunctioning projector.

Finally, Keith Redenbaugh gave his presentation. It consisted of a review of essentially all the scientific data the company had submitted to the FDA on the Flavr Savr tomato, with the exception of the risk analyses related to the potential for horizontal gene transfer to human gut or soil bacteria (see below). He had never described a Southern blot analysis in a speech before and made a valiant effort to explain the details of the molecular experiments we had carried out. He even mentioned specific restriction enzymes that had been utilized. Andrew Baum, president of Calgene's Oils Division, left the room during a particularly detailed account. John Callahan, continuing his efforts to lighten up the proceedings, noted that Andy must have been hungry, since he exited just after the enzyme Bgl2, pronounced "bagel 2," was mentioned.

But Keith's lack of molecular expertise was evident in his explanations of the DNA analyses, and he completely misidentified one of Bill Hiatt's protein experiments. The jokes I'd been hearing in the labs about whether Calgene would let the ball dribble through its legs at first base suddenly didn't seem very funny anymore. I was relieved to find out that Vic Knauf, Calgene's vice president of research, Don Helinski, director of the Center for Molecular Genetics at U.C. San Diego and a member of Calgene's Science Advisory Council and board of directors, and Bill Hiatt would all be present at the FDA meeting to field technical questions.

Tom Churchwell tried to promote an optimistic outlook on the upcoming meeting by reminding us that the FDA was not legally bound by its advisers' decisions, although the agency typically followed them. Since his knowledge was based largely on NutraSweet, whose approval process had turned into something of a public relations nightmare[2] (its management ultimately changed the name of the company to Benevia, from the Latin words for *good* and *way*), my concerns were not assuaged. So, despite Tom's assurance, I spent the days leading up to the Food Advisory Committee meeting working with Keith on his presentation of the DNA studies.

The day Bill and Keith were scheduled to leave for Washington, D.C., a couple of days before the FDA's Food Advisory Committee meeting, they came in to work early to finalize preparations. They were still going through a box of documents after 9 AM when I overheard Bill say, "We could use this . . . ," as I walked by Keith's office. Apparently they decided to keep their options completely open, since I saw Bill carrying the entire box down the hall a few minutes later just before they left for the airport. I couldn't help but wonder whether we were really ready for this momentous occasion. We weren't. Keith continued to be grilled by the rest of Calgene's entourage right up until the moment the meeting started.

The FDA Food Advisory Committee Meeting

Aside from Keith's presentation and some short introductory remarks by Don Emlay, the rest of Calgene's safety data were presented at the meeting by Julianne Lindemann, Ph.D. Julie had been working in Regulatory Affairs at Calgene with Keith and Don for about 18 months before that "unprecedented meeting of [the FDA's] Food Advisory Committee."[3] She was best known, however, for an infamous incident in the history of genetic engineering 7 years earlier. Donning a "moon suit," as was then required by the EPA, she sprayed ice-minus[*] *Pseudomonas* bacteria on field-grown strawberry plants for a company called Advanced Genetic Sciences, Inc., of Emeryville, California. Her picture, dressed in said suit and taken by unprotected photographers standing some 10 feet away, was printed in newspapers all over the world. Unlike her role in that episode, her participation in Calgene's historic event didn't involve bequeathing any garments to the Smithsonian Institution.[4]

On Thursday, April 7, Julie walked the meeting attendants through Calgene's estimates for the remote possibility that the *kan*[r] gene might be transferred from Flavr Savr tomatoes to bacteria in human or animal gut or agricultural soil. She emphasized the minuteness of the numbers Cathy Houck and I had calculated for the *kan*[r] advisory opinion document. She said that even our worst-case scenario, in which "1 new kanamycin-resistant [*Streptococcus*] bacteria would be produced for every 750 billion that are already present in the human GI tract," represented "11-plus orders of magnitude margin for error in which to conclude that such transformation does not constitute a significant risk factor." Similarly, she described the chances for soil bacteria to become kanamycin resistant via Flavr Savr tomato debris left in the field as "leaving us with 7 orders of magnitude [for *Bacillus* species and] . . . 10 orders of magnitude [for *Agrobacterium* species] margin for error to conclude that such transformation does not constitute a significant risk factor." Julie's presenta-

---

[*] Ice-minus was a term given to *Pseudomonas* bacteria that had had an "ice nucleation gene" deleted from their DNA through the use of genetic engineering. The idea was that the altered bacteria, when sprayed on crops, would render them frost-resistant.

tion made it difficult to conclude that the chances for transfer were anything other than insignificant.

She concluded by saying that the data she and Keith had presented were "merely representative of the safety documentation prepared by Calgene for the FLAVR SAVR tomato." For additional details, she suggested that the "public docket should be consulted." She stressed that the material Calgene had presented illustrated the "kind of reasoning and analysis that we think is sufficient to evaluate the safety of food products derived from genetically engineered plants."

If she had made her last statement even 1 day earlier, it might have elicited a debate over the sufficiency of Calgene's science. That's when some committee members had suggested, echoing a stand taken by Rebecca Goldburg and Environmental Defense, that additional experiments on horizontal gene transfer should be carried out. (Jeremy Rifkin and members of his organization did not attend the public meeting.)

But the FDA had successfully defended its position, and by Friday afternoon the Food Advisory Committee had been convinced. In the end, one attendant described the scene as a "love fest," especially in contrast to what he called the "bloodbath" that had greeted BST, Monsanto's genetically engineered hormone for dairy cows, during its approval process the previous fall.[5] As a bonus, the FDA declared that it saw no reason to limit the number of kan^r genes or the amount of APH(3')II protein in genetically engineered plant products, as Calgene had previously offered to do in its first submission to the agency.

The PR aspects of the meeting rubbed off favorably on Calgene, too. In addition to various newspaper reports, the Flavr Savr tomato was the focus of a story on "CNN Headline News" that Wednesday, and Friday morning Roger Salquist was interviewed on CNBC.

The following Tuesday, April 12, Vic Knauf shared his take-home message from the Food Advisory Committee meeting with Calgene's entire science staff:

> I know this is not news, but these data get uncommon scrutiny both scientific and otherwise. As we move on to new products, we absolutely need to do exquisitely clean science or some committee [member] with a scientific vocabulary could gum things up. The standards are tighter than publishing in *Nature* or [*Proceedings of the National Academy of Sciences* of the United States of America] where you only have to convince a couple of very intelligent reviewers. But let me assure you that Calgene Science sure came out of last week like a knight in shining armor.

At an all-employee meeting later that day, Roger said, "The event of last week may have been the single most important step in establishing this new industry." He said it had served as a "verification of Calgene science" and "you should all feel very proud." Finally, I thought to myself as I listened to the gushing going on around me, regulatory science at Calgene had earned a little respect.

But the news Roger had to report at that meeting was not all good. Although a crop of Flavr Savr tomatoes planted in Mexico had another 2 weeks' worth of fruit available for harvest, the chances of Calgene's receiving an official written blessing from the FDA in time to sell it were minimal to nonexistent. And the California crop wouldn't be available until the latter part of May. Julie Pear, unarguably Calgene's most outspoken employee of all time, then asked what most of us were wondering silently.

"So if FDA drags ass, will we go ahead and sell tomatoes anyway?"

Without hesitation, Roger replied, "Yes."

The Biggest Month in Calgene History

Roger's apparent willingness to abandon his nearly 5-year-old plan to obtain FDA approval (although not legally required to do so) before selling Flavr Savr tomatoes was brought on by Calgene's unhealthy financial condition. And that condition seemed even worse when the financials for the previous quarter were announced Friday, the thirteenth of May.[6] Although both the cotton and oil divisions were "performing according to plan,"[7] the bottom-line held a $9.8 million loss. The company was running so low on cash that it had registered 2 million new shares of stock with the SEC as part of a plan to raise $20 million through three private placements.[8]

In the local newspaper account, delay in obtaining approval from the FDA was blamed for the latest losses.[9] Calgene's own press release, however, reported that the financial results reflected "substantial losses associated with producing and marketing non–genetically engineered tomatoes."[10] And, although Tom Churchwell had officially resigned the previous Friday (with a nice e-mail send-off from Roger Salquist), the Calgene Fresh "scale-back" initiated a few months earlier to "reduce operating losses"[11] was still not quite complete.

On Monday, May 16, Rod Stacey set out to finish the job. He made Ken Moonie vice president of Midwest operations for Calgene Fresh. One of Ken's primary new responsibilities was to close the subsidiary's administrative facility in Illinois. Twenty-eight months and multiple tens of millions of dollars later, the Evanston era had come to an end.

But those dark financial times at Calgene preceded the dawn of a whole new agricultural age. Literally. The day after Ken's promotion, Keith Redenbaugh received the call from Jim Maryanski at the FDA. The approval letters for the Flavr Savr tomato and *kan*ʳ would be faxed to Calgene at 9 AM ET the following morning.

Keith was at work before 5:30 AM PT the next day, May 18. Sure enough, the formal notice from the FDA "supporting the safety of tomatoes grown from FLAVR SAVR seeds" rolled off the fax machine a couple of minutes after 6:00. Keith waited anxiously for the second letter, the agency's formal response to Calgene's FAP concerning the selectable marker gene, *kan*ʳ. It didn't come. Approval for Flavr Savr tomatoes without approval for the *kan*ʳ gene and the APH(3')II protein those tomatoes contained was practically useless. Roger and his response team, ready to spring into action, would be there any minute. Bearing bad news to Roger, especially this bad news, was about the last thing Keith wanted to do.

But by 6:29 AM, confirmation of the food additive regulation for *kan*ʳ had also come through, and Roger was sending a memo to the entire company that began, "The day we have all been waiting for has finally arrived . . . ." Most of Roger's note contained instructions about reporting any suspicious activities to the proper (in-house) authorities and dealing with the exceptionally high volume of telephone calls he expected. But he also specifically thanked Don Emlay and Keith Redenbaugh and invited the rest of us to join him "in patting [them] on the back for a job well done." He concluded by saying that it was a "big day in the history of Calgene, Calgene Fresh, and the food industry. Thank you all for helping to make it all possible." By 6:52 AM, the company's official press release announcing the FDA's decision had been issued.

Although one of Roger's goals for that special day was to "maintain the day-to-day business activities of the company," it quickly became obvious that day-to-day maintenance would be impossible for nearly everyone. A banner was placed on Calgene's front door that read "FDA Approval" in bold letters. (Incredibly, several of my scientific colleagues, who must stare at their feet when they walk, entered the building through that door without noticing the sign.) The Calgene TGIF event planned for that Friday was renamed the "Official Thank You FDA Golf Tournament." Even Roger couldn't keep his resolve. At noon he served champagne to the entire Calgene Davis staff while a local news crew recorded the celebration.

FDA approval of the Flavr Savr tomato made the front page of most U.S. newspapers the next day, and response to the news was almost uniformly positive. Financial analysts at firms like Merrill Lynch, spokespersons for Calgene's competitors, like DNA Plant Technology, and scientists from the FDA itself all touted various virtues of the Flavr Savr tomato. Even Rebecca Goldburg, senior scientist with Environmental Defense, admitted that the "FDA did a considerable review [of the Flavr Savr]."[12] Her organization made it clear, however, that its primary concern, as voiced by Environmental Defense attorney Douglas Hopkins, was not with Calgene's tomato, but with the fact that the FDA's process for dealing with genetically engineered foods was a "purely voluntary system where it is up to the companies to decide if a new food product is . . . to be reviewed."[13] With uncanny accuracy, Environmental Defense foresaw the next half decade, in which ag biotechnology would become "agriculture's most carefully cultivated secret."[14]

Jeremy Rifkin, on the other hand, marked May 18, 1994, as the beginning of his "tomato war."[15] He claimed that middle-class Americans were "moving in the direction of organic, healthy, sustainable foods" and the last thing they wanted to hear about was "gene-spliced tomatoes."[16] He vowed to "picket markets, hand out notices to consumers, and organize 'tomato dumpings' and boycotts" in protest wherever the Flavr Savr tomato was sold.[17] And, since his Pure Food Campaign had chapters nationwide by then, he had the ability to make good on his promise.

Roger Salquist was incensed with Jeremy Rifkin and his campaign. "Now they're trying to scare consumers," he told a reporter for the *San Francisco Chronicle*. "These people are parasites. They've been against technology since the development of the wheel."[18] In response to Rifkin's threat, Roger warned Calgene employees that "it is critical to our launch plans that you do not release any information regarding where we will sell the first FLAVR SAVR tomato." To the media, Roger revealed only that "limited quantities" of the Flavr Savr tomato would be available "in Northern California and the Midwest in the next 2 weeks."[19]

Roger's secret didn't have to be kept very long, however. The world's first genetically engineered whole food went on sale Saturday, May 21, 1994, 3 days after the FDA officially pronounced it safe. The "national rollout" took place in two grocery stores, the State Market IGA in Davis, California, and Carrot Top in Northbrook, Illinois.

In Davis, Flavr Savr tomatoes were displayed in a beautiful cart with a canvas awning. Each tomato carried a MacGregor's sticker, complete with farmer, watering can, and "GROWN FROM FLAVR SAVR SEEDS" spelled out in all caps. Bright-red, tomato-shaped brochures proclaiming "Summertime Taste . . . Year-Round!™" on their covers accompanied the real fruit. In addition to suggesting that MacGregor's tomatoes not be stored in the refrigerator, the point-of-purchase information briefly explained the genetic engineering process, including the use of the kanamycin-resistance selectable marker gene to produce Flavr Savr seeds. Consumers were invited to call 1-800-34TOMATO if they wanted still more information.

Although originally introduced at Carrot Top for $2.79 per pound, both stores soon carried MacGregor's tomatoes grown from Flavr Savr seeds for $1.99 per pound, about $0.70 a pound more than the prevailing premium tomatoes.[20] They sold like hotcakes. Bert Gee, the owner of State Market, resorted to limiting customers to two Flavr Savr tomatoes a day.[21]

Both stores were targeted by the Pure Food Campaign. Protesters in Davis threw tomatoes into a cardboard coffin in State Market's parking lot. But the size of the demonstrations, as well as the media's coverage of them, was minimal. Fewer than a dozen demonstrators participated in the Davis event, and it had no adverse impact on sales. In fact, Gee sold twice as many Flavr Savr tomatoes the day after the coffin incident than the day before.[22]

The celebration that had started with FDA approval the previous Wednesday continued at Calgene the Monday morning following the first sales. Even a rumor that it had cost the company $10 per pound to deliver its $2-per-pound tomatoes didn't seem to dampen anyone's spirits. Bill Hiatt prepared to party. He sent e-mail invitations to a gathering at his house commemorating the FDA's response and the launch of the Flavr Savr tomato. To my surprise, I got one.

In addition to the less than ideal way in which I had exited Bill's tomato program, he and I had been going head to head over a manuscript I wrote soon thereafter. The one-page paper documented the phenomenon that David Stalker had called "an act of God," the fact that the DNA Calgene's genetic engineers (and others) transferred to plants often consisted of more than what had previously been defined as "transferred" or T-DNA. Especially since the Food Advisory Committee had considered exact transfer of DNA an important issue during the public meeting in April, Bill was adamantly opposed to publishing the paper. Since the T-DNA issue could affect the work of not only industrial genetic engineers but also academic plant scientists, I was just as adamantly pushing for its publication. Fortunately for me, Roger Salquist and the rest of Calgene's executive staff took a proactive approach to (and consequently my side in) the controversy. To placate Bill, I was not allowed to use the word *tomato* in the text of the paper.

Science, the journal I submitted the manuscript to, has the unusual policy of requiring authors to include the names of colleagues who have reviewed the paper prior to its submission for publication. The political, social, and economic ramifications of this particular paper were such that it had been read, not only by Calgene's top scientists, but by most of its executive, regulatory, and legal staff and a few representatives of the company's biggest contract partners, such as Unilever, as well. I listed them all, starting with Roger Salquist.

Upon peer review, one referee found it "very important," "an outstanding contribution," but comments made by the other implied he or she didn't quite believe the results. Consequently, the paper was rejected. It was eventually published in *The Plant Cell,*[23] a journal with a circulation not nearly as large as that of *Science.* Still, the article did reach a worldwide audience. A member of the Ministry of Environment in The Netherlands, for example, wrote to say that he and other officials of the European Union were grateful for the information.

But Bill Hiatt and I put aside our professional differences to celebrate. It felt like old times (almost) as we spent most of his party watching a basketball game. When it came time for toasts, Don Emlay said that the "whole process from the start of the *kan*^{r} document to the last few questions asked by the FDA . . . [had been] great fun." Despite the festivities, his assessment of the previous 4 years brought the contrarian in me back up to the surface.

"I'm afraid you're rewriting history," I said. "This was a long, hard, difficult process. 'Fun' just does not describe it. It was more like having a baby. And now that the baby has finally been birthed, yes, it's very exhilarating." Bill, Don, and Keith politely agreed with the analogy. It's likely, however, that only a mother could really appreciate the comparison.

The Taste Tests

MacGregor's tomatoes were, of course, on the menu at Bill's party. As we gobbled up fresh Flavr Savr tomato salsa, Bill reminded us that "people are now eating our experiment." Most of Bill's guests, however, and a few select others, had been eating or at least tasting Flavr Savr tomatoes for years by then. And, although Roger Salquist claimed that "everybody who has tasted them says they are great"[24] (a statement that probably excluded him—he disliked tomatoes so much that he put off trying Calgene's triumph for years), the company's claim that its first genetically engineered product would save tomato flavor had always been controversial.

The results of Calgene's first "sensory evaluation," conducted in June 1990 and included in the original Flavr Savr tomato advisory opinion document filed with the FDA, were particularly disappointing. In a blind test, the judges, primarily Calgene employees, detected no significant differences in taste between Flavr Savr tomatoes picked pink and other tomatoes, genetically modified or not, that had been picked while still green. Not a very encouraging result when your business plan calls for high-volume handling of pink-stage tomatoes on the premise that they'll taste so much better than tomatoes picked green that you can charge a significant premium for them.

Those results, however, were easily rationalized. Because FDA approval to eat Flavr Savr tomatoes had not yet been obtained, those early "taste tests" had been of a "sip and spit" variety. Judges were allowed to smell, slosh around in their mouths, and chew fruit samples but not actually swallow them.

What's more, the sipping samples lacked locular material. Locules are the central compartments in a tomato harboring the slimy tissues that contain the fruit's seeds. Before Calgene received "non–plant pest status" from the USDA, Flavr Savr tomato seeds had to be "contained." And because tomato seeds are tough enough to pass intact through a person's digestive system (Charles M. Rick, the eminent U.C. Davis tomato geneticist, found, in fact, that the seeds of one wild tomato species, *Lycopersicon cheesmanii,* would not germinate at all until after they had traveled through a Galápagos turtle), allowing humans to swallow Flavr Savr tomato seeds in 1990 would have presented a considerable containment problem. Fruit locule contents were therefore eliminated from the first Flavr Savr tomato taste tests so that containment would not be an issue. However, because locules harbor not only the seeds but also the majority of the acid, some of the sugars, and therefore much of a tomato's taste, the design of those early tests was also fatally flawed.

After October 1992, when the USDA deregulated the Flavr Savr tomato, Calgene's taste tests became considerably more meaningful. The company provided intact MacGregor's tomatoes, grown from Flavr Savr seeds, to members of the media and let them design tests of their own. The resulting reviews were mixed.

A panel assembled by the *San Francisco Chronicle* found the Flavr Savr tomato nearly as good as end-of-the-season, organically grown fruit. Both tomatoes scored significantly below the panel's "excellent" rating, however.[25] Several people found freshly picked Flavr Savr tomatoes provided by the *Sacramento Bee* "significantly better than standard supermarket varieties." They were only willing to pay 10 or 15 cents more a pound for the taste improvement though, instead of the dollar or more per pound Calgene had been banking on.[26] And two chefs tasting for the *Washington Post* concluded that "MacGregor's is really no challenge to home-grown summer tomatoes."[27]

Michael Guillen, science correspondent for "Good Morning America," thoroughly enjoyed the Flavr Savr tomato he consumed during a broadcast, but other individual tasters were not as enthusiastic. Alice Waters, who had already publicly announced that she would not serve them,[28] said the Flavr Savr tomato she tasted was "not bad" but certainly not good enough for diners at her restaurant, Chez Panisse.[29] And Molly O'Neill, of the *New York Times* News Service, added insult to injury with her evaluation. She not only described its "character" as "too ambiguous . . . suggest[ing] more tomato than it actually delivers, taste-wise,"[30] but also, at least in some West Coast publications, incorrectly reported that the Flavr Savr tomato contained a "flounder gene . . . spliced into its chromosomes."[31] The flounder gene misconception persists as of this writing.

The Competition

Despite the less than glowing flavor reviews—partly due, no doubt, to the fact that its debut coincided with the summer tomato season—Flavr Savr tomatoes continued to sell exceptionally well. And Calgene had a big jump on its ag biotech competitors, at least in terms of getting genetically engineered tomatoes to market. Granted, the Freshworld Farms VineSweet tomato, a product of the collaboration between Meyer Tomatoes and DNA Plant Technology (DNAP), was already available and also selling for $1.99 per pound in May 1994. But DNAP's first truly genetically engineered tomato wouldn't be ready to market for nearly a year.[32] And Monsanto, like Calgene and DNAP also marketing conventionally developed, premium vine-ripe tomatoes in anticipation of selling genetically engineered fruit, wouldn't have its biotech version ready for 18 months or more.[33] Calgene had a momentary monopoly on sales of genetically engineered tomatoes.[34]

The company had its work cut out for it, to ensure that the moment would not be fleeting. The distribution systems of Calgene's competitors were extensive. Monsanto's Premium Ripe tomato was being tested in ten Eastern markets, and the Freshworld Farms tomato was already in ten supermarket chains, more than 1000 stores, across five states.[35] The number of stores selling Flavr Savr tomatoes, in contrast, was expected to increase from 2 to about 70 by the end of October 1994.[36]

In addition, as the adrenaline started to wear off, the truth came out behind that rumor about costs running $10 per pound for Flavr Savr tomatoes selling at $1.99 per pound. Margins on tomato sales were in fact significantly negative and in turn were draining Calgene's bottom line. In a meeting with his principal scientists during summer 1994, Roger Salquist mentioned the upcoming public release of the company's financial results for fiscal year 1994. As if an afterthought, he added, "Of course, they're horrible." And they certainly were. The company's net loss for the year that ended June 30 was $42.8 million, fully one-third of the total losses incurred over its entire 14-year history.[37]

Even worse, Calgene's fresh-market tomato business was expected to continue experiencing negative gross margins "at least through mid–fiscal 1995."[38] The company's strategy for achieving positive margins was to reduce its high unit costs for tomatoes, as summarized in its 1994 annual report:

Cost reductions will depend primarily on (1) tomatoes with the FLAVR SAVR gene [having] reduced spoilage; (2) the company achieving . . . increased crop yields, innovative production, packaging, handling, and distribution methods and . . . additional experience in the business; and (3) production and sales volumes reaching levels that would provide substantial economies of scale.[39]

Calgene was obviously still stuck on the double black diamond slope of the tomato business learning curve, and the end of the run seemed nowhere in sight.

Roger's response to the 1994 fiscal year results was rapid and multifaceted. He cut out coffee, started exercising more, and went back out on the road to "offset the quantitative crap with qualitative stuff" and raise more money. The most frequently asked question he got during his "dog and pony show" was, "When will you be profitable and why should I believe you?" His prediction that Calgene would be in the black during fiscal year 1996 must have been an acceptable answer, because a public offering of another 2 million shares of the company's common stock was completed on October 25. After expenses (including the tomatoes Roger had served to potential investors), the offering raised $16.4 million.[40] It was the company's sixth public financing and brought the total raised during its history to $270 million.

News of the successful public offering somewhat offset the company's first-quarter fiscal 1995 financial results, a net loss of another $9.5 million. The gross loss on net product sales of tomatoes, "substantially all" non–genetically engineered tomatoes, during the quarter had actually increased by $801,000.[41] That was more than enough for Roger. He vowed that the company would "not market any non–engineered tomatoes from [t]here on out."

Despite Calgene's financial straits, which could, after all, be blamed on the sale of traditional tomatoes, there was reason to remain optimistic at the end of calendar year 1994. Although supplies remained low, customer demand for Flavr Savr tomatoes was still high. What fruit was available continued to "fly off the shelf," and a waiting list of grocers wanting to stock MacGregor's tomatoes was established.[42]

The expansion of distribution went better than planned in October, and before Thanksgiving Calgene was delivering tomatoes grown from Flavr Savr seeds to 733 stores in the West and Midwest.[43] Huge ads devoted solely to MacGregor's tomatoes were run by the supermarket chains in the Sacramento area lucky enough to carry them.[44] Calgene's plan was "to supply up to 1000 stores by January 1995," because winter was viewed as the "best market" for its tomato,[45] and, it was hoped, an additional 1500 stores by June.[46] State Market's Burt Gee, who had received requests to ship Flavr Savr tomatoes to Alaska and the East Coast, sold gift packs—"four tomatoes in a box to give to friends and family"—over the holidays.[47] Although ultimately ephemeral, Calgene's momentary monopoly was certainly euphoric.

CHAPTER SEVEN

Afterbirth

On January 16, 1995, Bill Hiatt gave Calgene's Monday morning staff seminar, an update on activities at the Calgene Fresh subsidiary. He started with the good news. Demand for genetically engineered MacGregor's fruit was still very high. Therefore, Bill said, Calgene Fresh was having no trouble gaining coveted shelf space in grocery stores nationwide. As long as the company could supply premium-quality Flavr Savr tomatoes for those shelves, produce managers were making room for them.

Even transformed fruit that lacked MacGregor's quality, off-brand Flavr Savr tomatoes, had found a place in the market. They were being sold to fast-food outlets and other restaurant chains. Calgene's genetically engineered tomatoes were being served at places like Burger King and Denny's.

However, Bill explained, the supply of Flavr Savr tomatoes was going to be very limited throughout January, February, and March, a situation that also had been announced in a company press release the previous week. There were two main reasons for the shortage. One was an(other) act of God. Tropical Storm Gordon had wiped out most of the tomato (and pepper, cucumber, and strawberry) crops in southern and central Florida earlier that season. The other was an act of man(agement). Calgene Fresh had failed to plant a winter crop in Mexico.

But, Bill went on to say, by the time the next big crop of Flavr Savr tomatoes was ready to be harvested, a crop Calgene Fresh believed ample enough to satisfy customer demands,[1] two new distribution centers would be up and running to handle it. A 90,000–square foot leased facility in Immokalee, Florida, was scheduled to begin operations in April, and a 65,000–square foot facility under construction in Lake Park, Georgia, was expected to be on line in May. Both would be "equipped with the most advanced features of optical sorting, soft-handling equipment."[2] This "soft-touch" Durand Wayland equipment, was originally designed to sort and pack peaches. A similar system had been modified for tomatoes and installed in Calgene Fresh's first distribution center, on the south side of Chicago.

The January 10 press release had gone on to enthusiastically explain how the "combination of these two facilities will expand Calgene Fresh's ability to deliver a year-round supply of flavorful tomatoes to most major markets across the United States and Canada." Bill might have said that during his seminar as well, I don't know. I didn't retain much of the rest of his speech after he somewhat casually, especially for Bill, mentioned that the Flavr Savr gene wasn't helping the company's tomato business out the way they'd expected. Calgene Fresh had to handle vine-ripened Flavr Savr tomato varieties, it seemed, just as gently as other, conventionally developed, vine-ripened fruit.

I grappled with Bill's foreboding statement for a long time after his seminar. The press release did not explain why the company had neglected the all-important winter market by not planting in Mexico. I had just assumed that lack of access to Mexican growing operations like those run by Meyer Tomatoes, Inc., was the reason Calgene was doing only "trial plantings in Mexico."[3] But from Bill's talk and discussions I had later with Cathy Thome, Calgene Fresh's new tomato breeder, it was evident that Kanti Rawal's "Velcro" tomato varieties were not all the company had hoped they would be. Production-level experiments on Flavr Savr tomatoes, some of the more recent of which included the "Velcro" varieties, had been conducted for over 2 years by then, starting with that first disastrous truckload shipped from Mexico to Chicago in December 1992. The subsequent tests, I inferred from Bill's talk (and learned later through some of Ken Moonie's tales from the tomato trenches), had also failed to support the idea that the Flavr Savr gene could keep vine-ripened fruit firm enough to pack and transport like mature green tomatoes.

Danny Lopez, Calgene Fresh's new president and CEO, corroborated Bill's assessment in print. Danny had been hired, based on 18 years' experience in the fresh produce business, most of it with Dole Foods, to replace Tom Churchwell the previous July. He described the advantage of the Flavr Savr tomato this way: "Our technology has shown repeatedly that the breakdown from ripe to rotting is significantly slowed down. That's the benefit. Our technology works at the back end."[4] He went on to state that "technology from other people works on the other end,"[5] implying that the Flavr Savr gene did not work on that other critical, from green to ripe, end.

The story on which our story stock had staked its future didn't seem to be playing out properly. The Flavr Savr tomato hypothesis—the tentative assumption that because "antisensing" PG enzyme increased a tomato's shelf life it would also allow it to stay firmer while it ripened so that it could be picked "vine ripe" and still survive shipment to market—apparently wasn't panning out. And, even as we were just starting to grasp this possibility in-house (one scientist's grasp took the form of a hand-drawn tomato sticker posted in Calgene's main lab advertising "McGreenor tomatoes, engineered for no reason"), the investment community had already started smelling an unhappy ending.

The Calgene Fresh disclosure that it would have only limited supplies of Flavr Savr tomatoes during the first calendar quarter had thrown some stock analysts for a loop. Sano Shimoda, president of BioScience Securities in Orinda, California, said Calgene had "created a change in expectations,"[6] a great entrepreneurial sin, with its announcement. The price of Calgene stock, which had been steadily decreasing from $12 a share at the end of the previous August, consequently started to slide below $6.

Critics were also not pleased with the expansion of Calgene's infrastructure. They felt the company was "spending millions to produce a distribution network before proving it [could] make any money with its tomatoes."[7] Roger Salquist, referred to in one newspaper article as Calgene's "combative" (as opposed to the previously "colorful") CEO, was quick to point out that investors were overreacting. They "oughta be damn happy" about the addition of the two new packing-shipping facilities, he said.[8] Scaling up production and sales volumes to "levels that [would] provide substantial economies of scale" was all part of the company's plan to reduce costs and thereby make money on its tomatoes.[9] But, "When you're number one, people take pot shots," Roger was quoted as saying.[10]

Potshots were also being fired at the company in the form of "short sellers" of Calgene stock. (A short seller borrows shares of stock from a brokerage firm and sells them, hoping the stock's price drops and he can buy them back at the lower price before he has to pay back his loan.) Short interest in the stock increased soon after the press release of January 10 and continued to grow for months thereafter.[11]

The negative financial publicity in mid-January 1995 made it painfully clear that the hopeful prediction made by William O. Bullock, Jr., in his ag biotech forecast a few months before was not likely to come to fruition. Calgene and its tomato were certainly not warm enough to ignite the "flame that thaws the freeze [of biotech stocks] on Wall Street."[12]

Gloomy news was not only in evidence on the financial pages. Closed-door meetings were being held. Glimpses of surreptitious lists had been seen. We Calgene employees knew another internal doomsday was about to occur.

Calgeners understood, after experiencing multiple Calgene layoffs, that at one time or another everyone had been on a list, at least on a long one. This time, though, I knew I was particularly vulnerable. Upon leaving Calgene Fresh, I'd chosen a project as far away from product development as I could get, not a wise thing to do if you were interested in job security during a company's "shift from research to marketing."[13] And it had become obvious in recent months that my new project, an attempt to improve the yield and quality of cotton fiber, went against my new boss's philosophy of genetic engineering. David Stalker believed in engineering binary (all or none) traits, such as herbicide tolerance. Looking for incremental improvements in fiber length or strength was bound to be on his project chopping block. I knew I was in a good position to make the short list.

Sure enough, David showed up on my doorstep the Friday before the Monday that the cuts officially took place.

"Does everybody else who is affected already know?" I asked him as I let him in.

"You know why I'm here?" he asked incredulously. We drank a beer, and I nodded politely as he told me he'd literally flipped a coin to determine who would stay and who would go.

But the feeling that we were family at Calgene was especially strong, even at times like these. I felt that David was taking the whole thing harder than I was. (Unfortunately for him, it wasn't over yet. He took a call from Bill Hiatt that afternoon in my home regarding a final addition to The List.) When my husband got home we ordered pizza, and David brought his daughter over to join the party.

I had no regrets. "If I had it to do over," I told my Calgene colleagues on Monday, "I'd choose the same basic research project again." They were all supportive, and David said I could keep my office for as long as I wanted to. I came and went as I pleased for another 18 months before finally cleaning out my desk.

In addition to the layoff of 10 percent of its staff, Calgene's press release that day, January 30, announced the company's financial results for the previous quarter, another $5.6 million loss.[14] But the loss had been nearly twice as much, $10.8 million, during the corresponding quarter the previous year.[15] Gross tomato margins were still negative, but costs were definitely dropping.

And Calgene's "back end" technology still held an advantage for a vine-ripe tomato business (which explained Roger's response to a group from Cornell University who suggested that PG might not work exactly the way Calgene had assumed: "Who cares? I'm not trying to win the science fair next Friday. I'm selling a product"). Though they weren't firm enough at the shipping stage, Flavr Savr tomatoes were slow to rot once they reached their destination. The limited shelf life of ordinary vine-ripened tomatoes, only 3 to 4 days, was one reason why, in the mid-sixties, the fresh-market tomato industry had made the transition from handling vine-ripe fruit to primarily mature-green fruit operations.[16] Calgene Fresh could count on vine-ripened Flavr Savr tomatoes staying firmer than non–genetically engineered ripe fruit for an additional 7 to 10 days on the shelf,[17] a substantial improvement.

But, if the Flavr Savr gene didn't help on the "front end" by making it easier, more cost effective, to get vine-ripened tomatoes to market, then the company had to deal with all the other reasons the fresh-market tomato industry had given up on vine-ripes. And people with vastly more experience in the tomato industry than anyone at Calgene Fresh "felt that, compared with reforming the tomato business, genetic engineering [was] easy."[18] Sun World, "one of the top produce companies in the country, for quality and for quantity,"[19] had proven that point just a few months before. The company, which introduced the successful DiVine Ripe vine-ripened tomato in 1990, filed for bankruptcy protection in October 1994.[20]

The Sun World lesson was especially timely for Calgene Fresh. The *Los Angeles Times* reported that the "most expensive project to go bad [for Sun World] was one involving tomatoes in Mexico. Though fertilizer and other essentials cost the same . . . transportation, border-crossing, and labor costs were far higher than Sun World expected." "When you get done, the unit-per-acre cost is higher in Mexico than in the United States," Howard P. Marguleas, Sun World's chairman, was quoted as saying.[21] Based on Sun World's experience, Calgene Fresh's decision not to plant in Mexico, despite its negative effect on supplies of Flavr Savr tomatoes, had been a good one. The company was in enough financial difficulty without increasing its already troublesome costs by extending operations into a foreign country.

Calgene Fresh hadn't completely given up on the idea, though. A Mexican crop was still practically essential for providing a year-round supply of vine-ripe tomatoes in the United States. Cathy Thorne was therefore busy breeding the Flavr Savr gene the good old-fashioned way, into tomato varieties that were adapted to Mexico (and others adapted to the various U.S. tomato-growing regions). By March, Danny Lopez estimated that Calgene Fresh was 1 to 2 years away from commercial-level production in Mexico.[22] Granted, it would be difficult to supply Flavr Savr tomatoes year-round in the meantime, but, especially in light of the Sun World situation, Calgene Fresh had decided to wait until it could more readily afford to plant commercially in Mexico.

Danny Lopez also announced that March that Calgene Fresh was reexamining the whole practice of producing vine-ripe tomatoes.[23] Apparently he believed the company was still up to the challenge of "reforming the tomato business." "We are packing ourselves," he said, another reason the company was bringing state-of-the-art packing and distribution centers on line. But "soft-touch" packing equipment only went so far. Therefore, Calgene Fresh was also "looking at tests in which tomatoes are picked more gently . . . and . . . at new packaging."[24] Cathy Thorne was also breeding for firmness, especially on the "front end." (Who knew, perhaps, as Don Grierson's research group had found, the Flavr Savr gene might work much better if transferred into just the right tomato variety.) As a result of her work, Calgene Fresh had better tomato varieties, more efficient varieties, every month.[25] The goal was still to be in 2500 retail outlets by the end of June and to be profitable in fiscal 1996.[26] The company had its work cut out for it but believed it was on the right track in March 1995.

Then the Enzo patent case went to trial. The previous summer, Roger had told his senior science staff that the Enzo situation was "ugly, no other way to describe it." Things hadn't improved much in the meantime.

The case revolved around four key patent issues: enablement, inequitable conduct, obviousness, and prior art. Calgene claimed that, although the results of Masayori Inouye at the State University of New York (who had been issued three antisense patents subsequently licensed to Enzo Biochem, Inc.) preceded its own work, Inouye had not sufficiently enabled someone "skilled in the art" to use the invention in plants. Inouye "antisensed" three different genes, but only in a bacterium, *Escherichia coli*. Calgene argued that demonstrating the technique in a bacterial cell was a far cry from proving it would work in a plant cell. At least one anonymous biotech analyst agreed, saying, "It's difficult to imagine how a one-celled discovery of antisense gives you the right to all antisense in the world."[27]

The issue of inequitable conduct came up because Inouye had neglected to tell the Patent Office about a few forays into other cellular systems he had taken that had failed. He "tried to do his antisense with other genes in *E. coli*, two genes in yeast, four genes in animals, and he started a plant project, and they were all unsuccessful. And he never told the Patent Office about that."[28] (He had, on the other hand, published those failures but, apparently, one should specifically notify the Patent Office about these things.)

Despite Inouye's lack of success with his plant project, Enzo claimed that the Patent Office should never have issued Calgene's antisense patent in the first place, that antisense in plants was obvious in light of Inouye's other results. Calgene countered that the Patent Office was well aware of Inouye's work and issued its separate patent covering plants anyway.[29]

The fourth issue had to do with prior art. Calgene had cited previously published research, carried out at the Fred Hutchinson Cancer Research Center in Seattle, in its original patent application. In dispute was whether that study, carried out in Harold Weintraub's lab, or Inouye's work came first on a calendar basis. If Weintraub's work was first, "then Inouye's work could be viewed as obvious based on prior art."[30] Weintraub had filed his own patent application, and, covering all its bets, Calgene had obtained a nonexclusive license to it.[31]

But, subsidiary to the prior art issue itself was the biggest controversy in the entire case, if only in terms of the publicity it generated. Enzo alleged that scientific misconduct had occurred during the preparation of the Weintraub lab's work for publication and brought in the big guns to substantiate those allegations, Walter Gilbert, the Nobel laureate from Harvard. Gilbert testified that he spent 80 hours examining the laboratory notebooks of Jonathan Izant, the postdoctoral fellow who had been first author on the Weintraub manuscript. He concluded that Izant had "trimmed and cooked and forged the experiment," and therefore it was "untrustworthy as a whole."[32] Dr. Gilbert went on to describe "'trimming' research as smoothing irregularities to make the data look extremely accurate and precise; 'cooking' as retaining only those results that fit the theory and discarding others; and 'forging' as reporting research data that is invented."[33]

In response to these claims of fraud, the Fred Hutchinson Cancer Research Center conducted an internal investigation and concluded that Dr. Gilbert's allegations were "wholly without merit."[34] A spokesperson for the prestigious journal that had published the Weintraub work said that "Cell stands by the article. Nothing is being retracted."[35] Harold Weintraub, whom Roger Salquist had described as "completely on our side" a few months earlier, unfortunately could not testify on his own behalf. He died shortly before the trial began.

Then, as if the trial itself was not bad enough, the Wall Street Journal published an inaccurate attribution related to Gilbert's testimony more than 2 weeks after he gave it. The article started on the wrong foot with its title, "Expert Calls Calgene Research on Gene-Altering Method Flawed."[36] Walter Gilbert had, of course, not called Calgene's research anything, since he had only examined Izant's work, not Calgene's. The article went on to say that Calgene was "seeking its own patent for the process,"[37] when in fact Calgene had been issued its antisense patent 3 years earlier. The author concluded that losing the case might delay Calgene's "nationwide rollout of the Flavr Savr this summer and jeopardize its stated plans to become profitable in 1996."

Roger Salquist got personal. In a statement released the day the Journal article was published, he said, "The timing and biased nature of this article, in combination with a previous article by the same reporter, causes us to question his motivations." Furthermore, "This litigation will have absolutely no effect on the rollout of our MacGregor's premium vine-ripened tomatoes." (The judge's ruling wasn't expected for perhaps 18 months.[38]) The Journal printed a correction the following day.

The National Rollout of
Flavr Savr Tomatoes

The rollout began in May. Genetically engineered tomatoes were shipped to 1700 stores in New England, the mid-Atlantic states, the Midwest, the Pacific Northwest, and northern California.[39] Half-page advertisements dedicated solely to MacGregor's tomatoes reappeared in the *Sacramento Bee*. At Sacramento area Raley's and Bel Air markets, they were still selling for $1.99 per pound, but one store in Seattle was selling the "Calgene tomato" for $2.43 per pound, nearly twice as much as other, local tomatoes.[40]

The jump in price at the Seattle store may have been related to improved flavor in that crop of Flavr Savr tomatoes. "Gene Technology Success" was the title used to describe the results of a small taste test conducted by the *Seattle Times*: "Most of the 10 tasters—who weren't told what kind of tomatoes they were tasting—said the MacGregor variety had a sweeter, more 'tomatoey' taste than the conventional tomato."[41]

Even more encouraging on the flavor front was a tasting conducted with a group of particularly discerning tasters, owners and employees of high-end restaurants. It was another blind test, the Flavr Savr tomato competing with a field of five other tomato varieties, including a couple of beefsteak tomatoes and a Florida hothouse-grown fruit. Before the unanimous winner's identity was revealed, one taster said it had "great color, great smell . . . you can taste the dirt."[42] However, when Drew Nieporent, who owned several restaurants in New York with Robert DeNiro and Bill Murray, found out that he, along with the other four panelists, had judged Calgene's genetically engineered tomato as best, he responded, "No! No way. I can't believe it."[43] He had expected the Flavr Savr tomato to look like the "product of a scientific experiment . . . wan, waxy, and uniform in shape." It was, on the contrary, "big, beefy, dark-red, and sweet, the closest he could imagine to the summer Jersey tomato."[44]

Everything, except of course the bottom line, seemed to be looking up at Calgene. Viewing the financial picture, however, could not be avoided. The company had reported the net loss of another $4.46 million for the quarter that had ended March 31 (down from the $9.8 million it had lost during the corresponding quarter the previous year).[45] Even worse, despite and in the midst of the big rollout, it announced that "because of reduced yields from Florida tomato fields" the company's revenue would be lower and its losses higher than had been projected for the fourth quarter of the fiscal year.[46] The company claimed that the situation would not change its "projection for becoming profitable for the first time in fiscal 1996,"[47] but the short sellers could smell blood.

In an article published in *Barron's* the Monday following Calgene's projection-altering announcement, one San Francisco area biotech portfolio manager singled out Calgene as a good example of a stock to sell short. Mark Lampert found the company's stock overpriced, its story flawed, and the value of its intellectual property called into question by the Enzo litigation.[48] "Over the last 5 years, this company spent over $150 million of investors' money on a product, the super tomato, that doesn't have a place in the world. They can't make money selling tomatoes," he said. Lampert believed that the company was "facing a Herculean task to avoid bankruptcy" and that the stock would go to "zero or 50 cents."[49]

Roger Salquist countered that Lampert was "an admitted short-seller . . . trying to . . . drive the price of our stock down" but not, unfortunately, until after publication of the interview triggered the second-busiest trading day up to that point in the history of Calgene's stock.[50] Two million ninety-five thousand shares changed hands that day. (Average daily volumes had been about one-third that high in the previous 6 months.[51]) Only the 2,520,000 shares that changed hands the day the FDA officially approved the Flavr Savr tomato had been higher.

Calgene stock dropped 62.5 cents that black Monday. Trading volume slacked off, and the price was back up to $6.25 a couple of days later.[52] Things remained relatively quiet until the following Thursday, when trading got unusually heavy again. Some 1,274,000 shares were traded on June 1. This time, the stock closed 81 cents higher. The reason? That morning David Faber, a CNBC reporter, quoted a "reliable source" as saying that Monsanto was considering buying part or all of Calgene.[53] The rumor spread like wildfire. Reportedly, more than 700,000 shares changed hands between 9:20 and 10 AM.[54] Spokespeople for both companies had no comment.

The rumors persisted for weeks, during which time Calgene attempted to raise $10 million with a private stock offering in Europe.[55] According to Sano Shimoda, the analyst with BioScience Securities, Calgene "was under tremendous pressure" to raise money before the end of its fiscal year, June 30.[56] Two days before that deadline, with the private placement in Europe still incomplete, Calgene signed a letter of intent with Monsanto. An explosion in Calgene stock trading—the price went as high as $9.625—took place after the agreement was announced. Nearly 4 million shares changed hands. It closed at $8.25, up 50 cents.[57]

The First Monsanto Investment

Under the terms of the proposed deal, which required share-holder approval, Monsanto would gain 49.9 percent of Calgene for $30 million in cash, two lines of credit for up to $85 million, and Monsanto's 49.9 percent of Gargiulo L.P., the "largest fresh tomato grower, shipper, and packer in the United States."[58] (As outlined in the original agreement, Monsanto's intent was to finance completion of the Gargiulo acquisition and put the assets into Calgene.[59]) Calgene would also have access to a smorgasbord of Monsanto's intellectual property in fresh produce and plant oils.[60] In exchange for its $30 million, $10 million of which was advanced to Calgene immediately, Monsanto would get 30 million shares of newly issued Calgene stock.[61] Monsanto would also be given four of the nine seats on Calgene's board of directors.[62]

Stock analysts called it a tremendous strategic move for Calgene. "Overnight, Calgene will become the largest tomato producer in North America," explained André Garnet of A. G. Edwards in St. Louis. Not only large, Gargiulo had two areas of expertise in the tomato business thought to be key for Calgene's success: operational know-how and valuable tomato varieties, or germ plasm.[63] "Our problem has been having the right germ plasm providing high yield, resistance, and high flavor in each growing area," Roger Salquist said.[64] Because of Gargiulo's good reputation for breeding and growing tomatoes and "with the deep pockets of Monsanto behind it,"[65] Garnet declared that "all those problems are behind Calgene at this point."[66] Roger looked forward to "managing a business instead of running around to the stock market [for financing] every day."[67]

The original letter of intent also included a "standstill" provision. Monsanto could not make further equity purchases in Calgene until September 1998, a situation touted as virtually guaranteeing "Calgene's independence as a private enterprise for at least 3 years."[68] "This lets Calgene continue to be Calgene," Roger Salquist said.[69]

Calgene did indeed continue to be Calgene, and it wasn't pretty. I ran into Bill Hiatt the morning the company's complete financial results for fiscal year 1995 were published. He described them as another "disaster." The company had gone a total of $30.6 million into the red for the year. Bill found the Monsanto-Calgene alliance a "good match" but expressed doubts as to whether that alliance would be enough to save the company. "We got bogged down in a swamp of regulatory issues and variety development. The question is, can we pull this rig out of the swamp?" His was the (multi) million–dollar question.

But luck just didn't seem to be riding with the Calgene rig. The company called that summer's tomato-growing conditions "difficult" and blamed them for limited supplies of premium MacGregor's fruit. Prices for the increased number of nonpremium tomatoes Calgene Fresh was left with were especially low.[70] And, as if that wasn't enough bad luck for one season, the Unabomber used Calgene's return address when he mailed his treatise condemning modern technology to the *New York Times*.[71] Association with that kind of publicity the company just didn't need.

By the end of September and despite its new packing facilities in Florida and Georgia, Calgene was focusing "on selling most of its tomatoes west of the Rocky Mountains."[72] Roger cited germ plasm as the key Florida problem. "Our varieties just didn't perform well there."[73] There had, however, also been the "pinhooker" problem.

Because fruit on an individual tomato plant doesn't all mature at the same time, once any of the crop starts to ripen, vine-ripe growers often have to pick tomatoes every day to avoid letting them get too ripe to handle. The extremely high labor costs associated with these daily harvests are yet another reason the fresh-market tomato industry has turned to mature green fruit instead.[74] Gassed green fruit growers can get away with harvesting their plants just three times over the course of a season. Between harvests, tomatoes at the "breaker," turning, pink, or red-ripe stage of ripening and therefore past the point of interest for the gassed green industry, are picked by pinhookers. Traditionally, these free-lance pickers pay the grower perhaps $3 per 25-pound box for tomatoes the grower is not going to pack anyway. The pinhooker then sells the box to local restaurants or road-side stands for about $5.[75] It's a win-win situation, as long as you're in the gassed green tomato business.

If, however, you fancy yourself in the vine-ripe tomato business, pinhookers are not likely part of the solution to your gross negative margins. And so it was with Calgene Fresh. Without obtaining prior permission and on numerous occasions, pinhookers in Florida had picked what Ken Moonie, among others, viewed as the cream of the company's MacGregor's crop and made off with the harvest.

Whether in part due to the pinhooker problem or not, Calgene suspended its tomato-growing operations in Florida and Georgia during winter of 1995–1996. The company grew its tomatoes in Mexico and California instead that season.[76]

By the time the 49.9 percent investment deal with Monsanto was final, on March 31, 1996, Calgene was closing its books on the third quarter of its 1996 fiscal year. The company had experienced a few bright spots during the 9 months between the time it had signed the letter of intent with Monsanto and the day the deal became effective. It had started marketing its second and third genetically engineered products: BromoTol cottonseed and a canola oil modified to contain nearly 40 percent laurate, a key raw material used in the manufacture of soap, detergent, oleochemical, and personal care products.[77] The Enzo patent decision had come down. Not only did U.S. District Court Judge Joseph J. Farnan uphold Calgene's antisense patent, he also ruled Enzo's antisense patents were invalid, making the issue of whether Calgene had infringed on them moot.[78,*] And the United Kingdom's Ministry of Agriculture, Fisheries and Food had certified Flavr Savr tomatoes safe for consumption, the first such clearance of an unprocessed genetically engineered food anywhere in Europe.[79] Final approval to market MacGregor's tomatoes in the United Kingdom was not expected for another 2 years, but Calgene was already free to sell its tomatoes in Canada and Mexico based on approvals it had gained from those countries a year earlier.[80] Roger Salquist declared that the "verdict of consumers in the United States [was] overwhelmingly favorable toward MacGregor tomatoes . . . [and Calgene expected to deliver its] benefits to Europe as well when the regulatory process [was] complete."[81]

On the financial front, however, Calgene management's fresh-market tomato dream had become a nightmare. The company had suffered net losses of $10.4 million in the first quarter[82] and $5.7 million in the second quarter[83] of fiscal year 1996. But those losses were nothing compared to the ones recorded during its third quarter, the worst in its entire

---

* A federal appeals court later limited the district court's judgment of invalidity to the specific claims that had been asserted against Calgene in two of Enzo's patents. The remaining claims in those two patents and the entire third patent were left valid and standing (Enzo Biochem, Inc., press release, September 27, 1999, "Enzo Biochem reports federal appeals court decision in case concerning antisense patents").

history. The net loss for that one quarter was more than $76.9 million![84] (Newspaper accounts attributed $73.8 million of that to noncash losses related to Calgene's acquisition of Gargiulo.[85]) So much for the company's chances of becoming profitable in fiscal year 1996. In an interview that April, Roger Salquist admitted that, "when all [was] said and done, the biggest mistake was we tried to make [the tomato business] too big, too fast."[86]

Monsanto's Second Investment in Calgene

From that point on, all was essentially said and done. On July 31, 1996, Calgene announced Monsanto's plan to purchase another 4.7 percent of Calgene. The "standstill" provision in the original 49.9 percent Monsanto-Calgene letter of intent, among other items, had obviously undergone some changes. If the purchase was approved by Calgene's shareholders, the additional 6.25 million shares of stock would bring Monsanto's stake in the company to 54.6 percent and give it majority ownership. Roger Salquist resigned before the deal was reported in the papers.[87]

"My 12 years at Calgene have been the most exciting and rewarding years of my life, and I wouldn't have traded them for anything," Salquist said. "We single-handedly paved the way for genetically engineered foods, established the world's leading proprietary plant genetic engineering technology base, and built the finest team of people in the business."[88] Roger went out insisting that a good-tasting, bioengineered tomato with strong commercial potential was still possible.[89]

The deal was approved by Calgene shareholders in November. Monsanto paid $50 million, $8 per share for stock that had been trading at $4.875 before the deal was announced. "That eliminated some of the worst fears" among investors, one stock analyst, John McCamant, associate editor of the *Biotech Stock Newsletter,* said.[90] At least it hadn't been a fire sale. McCamant was later asked if he thought Monsanto would further increase its Calgene holdings. "At some point, they'll probably take the whole thing. That might be in 5 years or so."[91]

It turned out to be more like 12 weeks. On January 28, 1997, Monsanto pitched a proposal to Calgene's board that it purchase all of the approximately 30 million outstanding shares of Calgene stock for $7.25 a share.[92] John McCamant called Monsanto's proposal "cheap" and said that accepting it would be "stupid."[93] Monsanto subsequently increased its offer to $8 per share. A special committee composed of "disinterested" members of Calgene's board of directors unanimously approved the proposal on April Fool's Day.

The End of the Road for Calgene

Ironically, Calgene's last quarterly financial report was by far its best. It was even in the black! Based largely on Gargiulo tomato and Stoneville cottonseed sales, the company reported a net income of $4.7 million for the first 3 months of 1997.[94] Not for 10 years, since the 1987 prosperity that had spurred the hiring spree during which I came on board, had the company recorded its net income without parentheses. But it was too little too late. A few weeks later, Calgene ceased to exist as a publicly traded company.

Hendrik Verfaillie, a Monsanto executive vice president, said that acquiring the rest of Calgene's stock would "promote the closer working relationships and the greater sharing of technologies that are only possible with full ownership of the company."[95] The delivery of a similar message to Calgene employees was interrupted by the sound of the Calgene sign being jackhammered into oblivion. A spokesperson for Monsanto admitted that some job duplications would exist as a result of the acquisition but declined to speculate about layoffs at Calgene.[96] Some things at Calgene were destined not to change with the new ownership.

The Fate of Calgene's Flavr Savr Tomato

Monsanto—in part, perhaps, because of its firsthand knowledge (via Calgene as well as Gargiulo) of the kind of money that could be lost in the tomato business or its lack of experience with food products and in part, no doubt, because it had its own priorities for ag biotech product launches—deemphasized the Flavr Savr tomato. In fact, the staff at Calgene for all fresh-produce research was reduced to three people after Monsanto took over the company. Rick Sanders was one of them. By then, Rick and Bill Hiatt were the only people left at Monsanto's "Calgene campus" who had been part of the original Flavr Savr tomato team. In response to interest from Kirin Brewery Company of Tokyo,[97] Rick gathered a few more technical details in the lab about ag biotech's first fruit. Then, in 1997, he wrapped up his Flavr Savr tomato package and sent it off to Japan. To my knowledge, and appropriately enough, Rick was the last person at Calgene to work on the project. Kirin's plan was to sell Flavr Savr tomatoes in Japan by the turn of the century.[98]

There are many reasons why the Flavr Savr tomato was not successful. Public outcry at the fact that it was genetically engineered was not one of them. Almost without exception during the course of its brief commercial run, demand for the Flavr Savr tomato outdistanced supplies. And while the issue of whether the Flavr Savr gene is truly capable of saving tomato flavor remains debatable, anecdotal evidence from taste tests conducted over time indicated that the taste of MacGregor's tomatoes did improve as the Flavr Savr gene was bred into better-tasting tomato varieties.

Jeremy Rifkin, of the Pure Food Campaign, and members of various environmental advocacy groups had no real beef with Calgene's tomato either. Most agreed that Calgene had done a good job of demonstrating its safety, and they were especially pleased that the demonstration had been carried out voluntarily and publicly. Although not required to do so, Calgene had even labeled its tomato. In many ways, the company had exemplified what opponents of genetic engineering might call "politically correct biotech behavior." (Rifkin, for one, was still concerned about kan^r, the selectable marker gene that confers antibiotic resistance, however.[99])

Competition from other genetically engineered tomatoes also had nothing to do with the commercial failure of the Flavr Savr tomato. FreshWorld Farms' Endless Summer™ tomato and Monsanto's foray into the fresh-market biotech tomato business hadn't affected the rollout of Calgene's tomato at all. Those "stories," in fact, weren't particularly successful either. The one biotech tomato that did meet with some success was the processing tomato version of the Flavr Savr tomato that came out of Don Grierson's program at the University of Nottingham. Campbell Soup Company, which had opened the door for Calgene by granting it the right to use the Flavr Savr technology for fresh-market tomatoes, had licensed the same technology to Zeneca Plant Sciences (a division of ICI) and its partner, Petoseed Company, for the processing tomato market.[100] More than 1.8 million cans of tomato paste derived from genetically engineered tomatoes grown and processed in California were sold in Sainsbury's and Safeway stores in the U.K. from 1996 through mid-1999. Processing costs for preparing that paste were considerably lower because energy-requiring steps designed to eliminate PG enzyme had been minimized. Much of those savings were passed on to consumers. At 20 percent lower prices, the clearly labeled genetically engineered product initially outsold conventional tomato paste by 30 percent.[101] Because it was a processing tomato, however, Zeneca's success had no negative effect on Calgene's fresh-market tomato business. (During the latter half of 1998, when the backlash against genetically modified foods started brewing, sales of Zeneca's tomato paste dropped off dramatically, and the following summer the paste was pulled from the shelves. As I write this, my friends in the business tell me that there are no genetically engineered tomatoes currently for retail sale in the United States or Europe.)

There were two main metaphorical characters in the Flavr Savr tomato story. The genetic engineering character was solid. The other character, fresh-market tomato business experience, was weak, poorly established. Calgene's lack of expertise in the business it counted on vertically integrating was high on the list of reasons why the Flavr Savr tomato failed. To be fair, the company did experience its share of natural (tropical storms, record heat waves, hurricanes) and "unnatural" (pinhookers, Unabombers) disasters in its relatively short lifetime. But to be realistic, those disasters are all (with the possible exception of the Unabomber) just part of the risky business of agriculture. In retrospect, perhaps

Calgene had no business in that business. Maybe an enterprise like David Stalker's "gene boutique" might have been the better, although smaller, way to go. Whether Wall Street would have been satisfied with Boutique à la Calgene we will never know. What we do know is that, in the end, Calgene was not up to the job of reforming the fresh-market tomato business.

Of course, maybe the company could have put together the necessary expertise, could have survived some bad luck, could even have made money, if only the Flavr Savr tomato hypothesis, the plot of this story, had played out perfectly. But, alas, most tomato varieties were not firmer as they ripened on the vine after having been genetically engineered with the Flavr Savr gene. The few that were firmer on the "front end" still could not withstand the handling and shipping, even "soft-touch" handling and shipping, necessary for a large-scale vine-ripe tomato business. And, perhaps the biggest disappointment of all, because tomatoes had been bred for toughness and with no regard for flavor for so long, most commercially available tomatoes, even when allowed to ripen on the vine, still had mediocre tomato flavor. When the dream sequence had faded and harsh reality had come back into sharp focus, the benefit of the Flavr Savr gene was still on the "back end." It slowed the rotting process in ripe tomatoes and gave them significantly longer shelf life. And, because it was a back-end benefit, the Flavr Savr gene was, at best, of marginal value to a fresh-market tomato business. Calgene, with minimal experience in that business, could not turn a profit on its marginal, albeit novel, product.

The flagship product of the ag biotech industry, the Flavr Savr tomato, sank. Before it disappeared beneath the waves, however, it played a monumental part in establishing that new industry, a part that was still being acknowledged by Jim Maryanski during the FDA's public meeting in Oakland in late 1999. The Flavr Savr tomato had opened the floodgates for the rest of the ag biotech fleet. Virus-resistant squash, New Leaf™ potatoes, Roundup Ready® soybeans, Bt corn, and other genetically engineered food products were then free to sail out into the marketplace in its wake.

The Aftermath

On June 24, 2000, Roger Salquist and his wife, Claudia, gathered a group of former employees in their home for a Calgene reunion. Although Ray Sheehy, Matt Kramer, Don Emlay, David Stalker, and I no longer worked at Monsanto's "Calgene campus," we all attended the party anyway. Many other Calgene scientists, some of whom had been laid off as far back as 1988, others as recently as a few weeks before the party, and a few of whom still worked for Pharmacia, Monsanto's then owner, were also there. (Not surprisingly, Bill Hiatt, one of the Pharmacia few, did not attend.) Norm Goldfarb, a founder of the company, its first chairman of the board, and long-time entrepreneur, told the crowd of about 60 that, of the many companies he'd started since Calgene, none had the same family feel. It was easy to agree with Norm and Roger that evening. We were all still part of the Calgene family.

As at any family reunion, people spent time catching up with each other, describing their new jobs and their kids (I brought pictures of mine), and sharing old memories. Professor Ray Valentine, Calgene founder on the science side, retold the tale of how President Richard Nixon personally intervened to secure a federal grant for his lab, a grant without which Ray's research would have lapsed and Calgene would never have been born. (I had first heard the story in Calgene's main lab on the day of Nixon's death; Ray had taken an oath not to repeat it until that time.) Calgene paraphernalia, like hats and shirts, were also passed around and worthless stock certificates laughed over. Several people gave me tidbits for this book. Everyone, including Roger's dog, Enzo, seemed to have a good time. It was a low-key, friendly affair—except when I broached the subject of the public debate over genetically engineered foods.

That's when I encountered enthusiastic renditions of the same arguments that people associated with the ag biotech industry had been making for at least a dozen years: "The public doesn't understand the technology," "It's just an extension of traditional breeding," and "Diabetics don't care if their insulin is genetically engineered, so why should the public care if their food is?" I was surprised, especially about the insulin comment, and reminded its deliverer of the reasoning we'd accepted at Calgene from the start: sick people are willing to take a much higher risk for their health than healthy people are willing to take with their food. Genetically engineered insulin is also labeled as such. Theoretically, anyway, a diabetic can weigh the benefits against any risks (perceived or otherwise) and make a decision accordingly. Consumers can't make similar choices when it comes to genetically engineered foods in U.S. grocery stores unless those foods are labeled.

The topic of labeling biotech foods provoked an even more emphatic response from my former Calgene colleagues. We didn't discuss the specifics, that proponents of ag biotech believed labels would "add significantly to production costs"[1] and implied "that the buyer needs to be warned."[2] But one industrial scientist pointed to her own arm and declared, "You might as well label me, too."

I wondered whether the agitation of these scientific party-goers on the subject of public acceptance was related to a sabotage attempt at the Calgene greenhouses by an anti-biotech group just a few days earlier. Reclaim the Seeds had said that the attempt was part of its "Operation Cremation Monsanto" and had been carried out "in solidarity with . . . people around the world who have resisted [Monsanto's] agricultural biotechnology program" and also because Calgene had created the Flavr Savr tomato.[3] The attempted break-in had been the first such action at Monsanto's Calgene campus. This kind of vandalism had been occurring in India, France, and Great Britain since late 1998 and had, starting in summer 1999, spread to companies and some universities (including U.C. Davis) engaged in biotech research across the United States. The claim of Reclaim the Seeds was that Monsanto conducted "secret operation[s]."[4] Some anti-biotech groups were responding in kind, only their operations were illegal as well as secret.

But, if my assumption was correct and the intensity of my former colleagues' remarks was related to the recent guerrilla attack, the conclusion they drew about the public outcry over genetically engineered foods seemed even more surprising. The spring planting of genetically engineered crops, they said, was complete. And, as they had done the previous year to the tune of nearly 100 million acres' worth, many farmers, the customers of companies like Monsanto, had chosen Roundup Ready soybean, Bt cotton, and other genetically engineered crop seeds to sow. Therefore, the anti-biotech food crusade—consisting of only a few thousand environmentalists anyway, these scientists and others in the ag biotech industry believed[5]—had proven ineffective and the entire controversy would soon blow over.

I was taken aback. The FDA had received some 35,000 comments from the public in response to its meetings on the regulation of genetically engineered foods held the previous November and December. Since those meetings, the *New York Times* had run a series of front-page articles on biotech crops, one called "Redesigning Nature: The Battle for Public Opinion." Public demonstrations against ag biotech had been staged, not only at the three FDA meetings, but also at the World Trade Organization meeting in Seattle, the U.N.-backed summit to forge a biodiversity safety protocol in Montreal, annual shareholder meetings for various companies, such as Safeway,[6] and some biotechnology conferences.[7] Some companies[8]—Gerber (a division of Novartis), Heinz, Frito-Lay, McDonald's, Kirin, Sapporo, and even Iams among them—had responded to public sentiment on the subject by eliminating genetically engineered ingredients from their products. And, likely the main reason ag biotechnologists worried about American farmers' seed choices for 2000 in the first place, there had been the statement released by Archer Daniels Midland (ADM) the previous summer. As one of the world's largest processors of oilseeds, corn, and wheat, ADM had caused quite a stir in agricultural circles when it encouraged its suppliers "to segregate non–genetically enhanced crops" in order to "produce products that our customers will purchase."[9] One marketer of genetically engineered soy and other seeds referred to the resulting ruckus as "the Y2K of agriculture."[10] The August 31, 1999, statement still represented ADM's official policy on the subject at the time of the Calgene reunion (and continues to do so as I write this).

So, contrary to the stand taken by a few Ph.D.-level scientists at Roger's reunion, the public debate over genetically engineered foods appeared far from over. In fact, after sending the FDA only a couple of dozen comments regarding the approval process for the Flavr Savr tomato, the general public in the United States appeared to be making up for lost time. In Europe and other parts of the world, of course, the debate over these products and the process by which they were wrought was also still raging. The royal family in Great Britain, for example, was split on the topic. Prince Charles was against, the Duke of Edinburgh for, and, after initially indicating that "fundamentally" genetic engineering was "much the same thing" as selective breeding and it was, therefore, "a bit cheeky to suddenly get nervous about it," Princess Anne concluded that "the jury was still out."[11]

After a moment of stunned silence, I therefore informed my scientific acquaintances that they were in denial. Another former colleague who'd participated in the conversation seconded my suggestion, and there were others at the party who were also more realistic about the public relations situation. (A multipage article entitled "Biotech Backlash" on the front page of the *Sacramento Bee* the next morning supported the realists' assessment as well.[12]) The denial incident, however, convinced me of a suspicion I'd had since witnessing the FDA meeting in Oakland the previous December. An inability on the part of at least some members of the pro-biotech scientific community to face and deal with the facts on the subject was contributing to the polarization of the debate.

Nearly every scientist, industrial representative, or U.S. federal regulator who has defended the use of biotechnology in agriculture, for example, has started with the notion that genetic engineering is an extension of traditional breeding. We at Calgene used the argument in our original requests of the FDA back in the early 1990s. But, despite the fact that this idea has long been the consensus among scientists,[13] at least among those practicing genetic engineering, it is not an established scientific fact. It is opinion. Granted, it is an opinion that many people, especially molecular biologists, might share for, say, the case of the Flavr Savr gene in the Flavr Savr tomato. "Antisensing" the tomato polygalacturonase (PG) gene using genetic engineering results in what is essentially a PG gene–specific mutant of tomato not that different from PG mutants created using traditional genetic methods. Fewer people, however, would agree that the practice of inserting the kan^r gene from *Escherichia coli* bacteria into tomato plants is on a continuum of traditional genetic modification techniques. The difference is merely a matter of opinion. And, since many anti-biotech Americans have indicated that they do not share that opinion, I question the value of clinging to the "extension of traditional breeding" mantra for either the purpose of defending this new technology or as a cornerstone of "science-based" U.S. regulatory policy.

Rather than personal opinion, the scientific community should give the public facts, hard facts: the results of studies that indicate these foods are safe to eat and that growing them on a large scale will not cause environmental damage. Scientists and regulators throughout the ag biotech industry agree that more public education about genetic engineering research is necessary, but, thus far, few have provided much information beyond how the technology works and what wondrous things might be done with it. Such minimal public education efforts, even combined with $50 million advertising campaigns,[14] will not get the job done. And simply proclaiming that "these foods are safe and there is no scientific evidence to the contrary"[15] is not the same as saying "extensive tests have been conducted and here are the results." In fact, without further elaboration, "no scientific evidence to the contrary" could be construed as "no scientific evidence, period."

Deficiencies in relaying factual reassurance about the technology have also been confounded by the fact that most of the few ag biotech–related scientific studies covered in the popular media have made the public more, not less, nervous about the safety of so-called GM foods. And, frankly, reports of super weeds and pollen that can kill monarch butterfly larvae and taco shells contaminated with corn not approved for human consumption have made me more concerned, too. These are the very kinds of problems that groups like Environmental Defense have been warning us about for more than a decade.

Some scientists responded to especially the monarch study by nit-picking about controls and downplaying its meaning for the "real world."[16] (It is worth noting that papers reporting both herbicide-resistant weed development and monarch larvae deaths were published in *Nature*, arguably the world's most prestigious scientific journal.) Personally, I found another matter that was worrisome. Bt, the insecticide produced by the bacterium *Bacillus thuringiensis*, is famous for killing lepidopteran larvae. Monarch caterpillars are lepidopteran larvae. Why weren't the types of studies initiated at Cornell University carried out before Bt corn was released on a commercial scale and grown on 20 million acres?[17] A variation of the same question applies to super weeds. Canola is notoriously promiscuous. There are lots of *Brassica* crop plants and some weeds it is known to genetically cross with. Calgene said as much in its first request for an advisory opinion from the FDA. Why weren't thorough studies of the potential for transfer of herbicide-resistance from canola to sexually compatible weeds conducted before a genetically engineered herbicide-resistant oilseed canola was ever put on the market?[18]

Living up to the standard of contrariness I set at Calgene, I also wondered why herbicide-resistance genes were being genetically engineered into canola in the first place. In light of the super weed potential, some companies (Pioneer Hi-bred, for one) chose not to commercialize herbicide-resistant canola at all. And why is Bt toxin in the pollen of Bt corn plants and the tubers of Bt potatoes anyway? European corn borers and Colorado potato beetles, the insects these plants were designed to be protected against, do not normally eat corn pollen or potato tubers, respectively. Granted, it would have taken considerably longer to utilize the precision of plant biotechnology to produce products that expressed the Bt toxin gene only in the specific plant organs that are attacked by that plant's pest or pests, but it could have been done.

The good news is that the public's interest in genetically engineered foods has spawned additional scientific research on ag biotech safety issues. The Cornell butterfly report was the first of its kind. A dozen or more labs have now joined the effort to define the risk Bt corn may present to monarchs and other beneficial lepidopteran insects.[19] Risk assessment is also becoming increasingly multidisciplinary. Instead of genetic engineers, who are molecular biologists, extrapolating from highly controlled lab and field trial conditions to commercial production, ecologists, entomologists, population geneticists, and environmental scientists are getting involved. A multidisciplinary approach to safety assessment was recommended in the original International Food Biotechnology Council (IFBC) report, the 400-page "navigational chart for both companies and federal regulatory agencies," back in 1990.[20] Some concerned ecologists felt that that approach was still underutilized in 1999, however.[21] And there likely will be many more studies undertaken in response to a report from the National Research Council of the U.S. National Academy of Sciences. The report, released in April 2000, called for "aggressive research" into the monarch butterfly situation, insect adaptation to pesticides such as Bt, the potential for human allergens in biotech foods, and other questions. It also urged the USDA, FDA, and EPA to fine-tune and better coordinate the regulation of genetically engineered foods.[22]

The need for better coordination among U.S. regulatory agencies is best illustrated by the Bt loophole. (For an excellent explanation of this loophole and the Bt food situation in general, I recommend Michael Pollan's article "Playing God in the Garden."[23]) Because Bt is a pesticide and pesticides are exempt from FDA regulation, the FDA does not regulate genetically engineered Bt foods. The EPA does. And as long as Bt levels in Bt foods are below a human "tolerance" level established by feeding Bt, in a bacterial form used as a pesticidal spray, to animals, those foods can receive EPA's green light. As an added catch-22, although Bt sprays carry warning labels, the EPA cannot label Bt corn or Bt potatoes because the FDA has sole jurisdiction over labeling foods (except meats and poultry, which come under the jurisdiction of the USDA). But, even if it wanted to, the FDA could not legally label Bt foods. The FDA is specifically barred from including any information about pesticides on food labels.[24]

In light of this loophole, the EPA began rethinking its approach to regulating genetically engineered plants in spring 1999.[25] Agriculture Secretary Dan Glickman also solicited advice in preparation for fine-tuning USDA policy. In March 2000 he put together a 38-member committee to review the USDA's role, not only in regulating biotech crops, but also in developing technologies like the one dubbed "terminator," which renders seeds sterile so that farmers cannot save seed stock from their crops but must repurchase seeds every year. (Although Monsanto backed away from using the terminator seed sterilization technique as a way of protecting its biotech investment, "the USDA has thus far refused to relinquish its patent on the process."[26]) And the Clinton administration, which wanted more federal oversight of genetically engineered foods and the details of that oversight to be more transparent to the public, came up with its own plan to affect the way that both the USDA and the FDA regulate biotech foods.[27]

At this juncture in the regulation of agricultural biotechnology it is, I believe, especially helpful to take another hard look at the approval process for the Flavr Savr tomato. I hope this book has demonstrated that the FDA did an especially thorough job of examining Calgene's safety data. For 4 years the agency put the company through the ringer (or at least that's how it felt to me), asking for additional information as early and often as it deemed necessary. David Kessler, FDA commissioner during the process, said, "Every safety aspect was examined in exquisite detail."[28]

The consequence of the FDA's exhaustive review, the agency's determination that the Flavr Savr tomato was as safe as other conventionally produced tomatoes, was a logical conclusion drawn from the data Calgene provided (or so I believe—readers of this book can decide for themselves). However, the FDA also concluded, in conjunction with its Food Advisory Committee, that subsequent genetically engineered products would not require similarly extensive reviews. Formal approval from the FDA for subsequent products was in fact not necessary at all. Instead, the agency put a voluntary consultation process in place. The test case of the Flavr Savr tomato, in my opinion, did not support this more general conclusion. Calgene's tomato should not serve as a safety standard for this new industry. No single genetically engineered product should.

Julie Lindemann, speaking in reference to the 1990 International Food Biotechnology Council report she helped prepare,[29] seemed to agree. The report's decision tree, sanctioned by the council's 30 member companies from the biotech and food-processing industries, was "not a laundry list," Dr. Lindemann said. It would be "very difficult to specify" the exact criteria necessary for making judgments about individual products. Those criteria were bound to be "crop specific."[30] I couldn't agree more. Safety assessment of these products needs to be carried out on a case-by-case basis.

What's more, the Flavr Savr tomato makes an especially inappropriate safety standard. If all genetically engineered foods were produced by antisensing a gene normally expressed in that food, as was done in the case of the Flavr Savr tomato, then minimally scrutinizing new genetically engineered food products might be reasonable. (An example of minimal scrutiny, a dossier filed recently for a genetically engineered linseed oil, which "appeared to address all [the FDA's] concerns," was only five pages long.[31]) Antisensing a plant gene involves the addition of only a piece (or a few pieces) of DNA to a plant. The source of the added DNA is, preferably, a member of the same species destined to undergo the transformation-regeneration process. No foreign protein product (or increased amount of an endogenous protein product, for that matter) is added to a genetically engineered plant produced using the antisense method.

But the genetically engineered traits in nearly all of the approximately 40 genetically engineered crops that have entered commercial production since the Flavr Savr tomato[32] have not been produced by simply shutting down an endogenous plant gene using antisense. Genes from bacteria, viruses, or other plants—and, yes, even a fish, although, as far as I know, the tomato harboring that gene was never commercialized—the expression of which results in the production of proteins foreign to the host plant, have been added to essentially all these foods. I would not put these products into the same innocuous category as the Flavr Savr tomato, unless . . . the *kan*[r] gene in that tomato is also taken into consideration. The *kan*[r] gene puts the Flavr Savr tomato, and the whole host of other genetically engineered foods containing *kan*[r] as well, into an entirely different regulatory category. At least the way I see it.

When it came to the selectable marker gene, *kan*[r], the FDA gave Calgene relatively conservative advice, bureaucratically speaking anyway. APH(3')II, the product of *kan*[r] gene expression, was in the end regulated as a food additive by the FDA, at least in tomatoes, cotton, and canola. I agree that APH(3')II is a food additive and have ever since Calgene's first regulatory team meeting in August 1990. APH(3')II is a protein not normally found in (as part of) the foods we eat, whether those foods were produced using genetic engineering or otherwise. When it is added to genetically engineered foods, it is therefore a food additive and should be regulated accordingly.

The FDA's appropriately conservative stance on *kan*[r], however, begs the question: if APH(3')II is a food additive, then what about Bt proteins and viral coat proteins and other proteins that were not part of our foods until they were engineered into them? Why aren't those proteins food additives too? And, if APH(3')II is a food additive in tomatoes, cotton, and canola, what about all the other genetically engineered crop plants produced using *kan*^r as a "processing aid"? It seems to me that those foods also contain at least one, and that one an "official," food additive.

Working from the logical premise that these engineered foreign proteins added to our food are food additives begs another question: shouldn't these food additives be listed on U.S. food labels? As I write this, the labeling of genetically engineered food products is on a purely voluntary basis, the significant exceptions being foods with substantial changes in nutritional content or containing suspected allergens. However, U.S. Senator Barbara Boxer, of California, introduced legislation in February 2000 that would require labels on products that contain genetically engineered ingredients. A similar bill was introduced in the U.S. House of Representatives. Boxer's suggested biotech food label would read: "This product contains a genetically engineered material or was produced with a genetically engineered material."[33] Grocers, food processors, and ag-biotech industrialists, including those I spoke to at the Calgene reunion, are opposed to the kind of mandatory, "blanket labeling"[34] described in Senator Boxer's bill. So am I.

Blanket labeling is too ambiguous. Since I am opposed to neither technological innovations generally nor agricultural biotechnology particularly, I need more information. I want the ability to select Flavr Savr tomatoes or Zeneca's tomato paste to feed to my family, for example, but to monitor New Leaf potatoes until a "new and improved" version lacking Bt toxin in its tubers is available. A general label like Boxer's doesn't give me enough information to make that choice. Including food additives like "Bt protein" and/or "APH(3')II" on food labels would. For cases in which uncertainty exists as to whether a food contains a particular genetically engineered food additive, the "may contain" strategy, already frequently used for other food ingredients, could be utilized.

Of course, it would take an incredibly informed public to (1) recognize and (2) evaluate these genetically engineered food additives. Even with 15 years of experience in the field of plant molecular biology, I probably don't know all the proteins added to our foods through genetic engineering by name. But dealing with this dilemma is a perfect opportunity for the ag biotech industry, which has long lamented the lack of consumer education on this technology, to bridge that information gap. And the Flavr Savr tomato serves as an excellent example of how to build the bridge. The tomato-shaped point-of-purchase brochure that accompanied Mac-Gregor's tomatoes not only explained genetic engineering, antisense technology, and Calgene's *kan*ʳ "processing aid" succinctly and in simple terms but also provided consumers with an 800 number "for more information about MacGregor's tomatoes grown from Flavr Savr seeds." Perhaps not many consumers will go to the trouble to call or visit a Web site or write a letter to ask about a food additive they don't recognize on a label. (In June 2000, Kraft Foods was receiving more calls about aspartame, the generic name for NutraSweet, than about genetically engineered foods.[35,*,36]) That's not the point. Those who are interested in finding out more about an additive could do so. That's almost infinitely more choice than U.S. consumers have now when it comes to genetically engineered foods.

The Clinton administration's proposed plan called for blanket-type labels, indicating either "free of" or "containing" gene-spliced ingredients.[37] It also called for that label-

---

* This situation changed dramatically, however, when StarLink™ corn, lacking government approval for human consumption, was identified the following September in Taco Bell–brand taco shells marketed by Kraft Foods.

ing to be voluntary. Although some ag biotech proponents believe that "most [genetically engineered products under development now] will be labeled voluntarily by the company,"[38] the industry's track record and the observation that "most of the food industry opposes labeling, . . . saying it might alarm consumers,"[39] don't support that belief. A portion of the ag biotech industry is against the biotech-free label also "because it might stigmatize or imply superiority over genetically engineered foods"[40] and "there is no system in the United States to segregate GM from non-GM" foods.[41] Validation that products are free of engineered ingredients could also be unwieldy and costly. Few associated with this new technology, it seems, are inclined to voluntarily label much of anything.

But some companies are finding ways to separate genetically engineered from non–genetically engineered ingredients in their products and are therefore poised to do so. Frito-Lay, one of America's top snack food makers, simply went to the source and told its contract farmers not to grow genetically engineered corn.[42] Kellogg, too, has been removing genetically altered ingredients from its products, although only in those bound for Europe and Australia, destinations where mandatory labeling for genetically engineered ingredients is either on the books or in the works. It has thus far declined to do so for its North American customers.[43] Apparently, companies can segregate genetically engineered crops (or food ingredients) from traditionally produced crops (or food ingredients) in order to correctly label foods as "genetically engineered" or "GMO-free," but some will require legal encouragement to do so.

The Clinton administration's plan (never finalized) also proposed changing the FDA's voluntary consultation process into a mandatory notification procedure, one in which the agency would make a company's safety assessment studies available to the public, as was done for the Flavr Savr tomato. That move would be a good first step the new Bush adminstration could take toward increasing consumer confidence in these controversial foods.[44] Calling for voluntary rather than mandatory labeling of these foods would not. Big corporations have choices when it comes to genetically engineered crops. Farmers can weigh the premium cost of genetically engineered seeds against the benefits associated with reduced tillage and fewer pesticide applications, the likelihood of being able to sell their product after harvest, and even concerns for monarch butterflies before they make a purchase. Farmers chose to plant more biotech soybeans and cotton in 2000, for example, but 20 percent less gene-altered corn.[45] U.S. consumers cannot make similar choices. It's time for our government to reexamine the FDA's APH(3')II precedent, reconsider its position that "genetically engineered ingredients are not inherently a food additive,"[46] and revisit Environmental Defense's 1991 proposal for utilizing the Food Additives Amendment of the Food, Drug, and Cosmetic Act to regulate biotech foods.[47]

As a society, we've been bad parents to this biotech baby for too long. The ag biotech industry, the baby's proud papa, has spent too much time bragging about what the child might be when it grows up and not enough time dealing with regulatory studies, the necessary dirty work that accompanies the joy and excitement of parenting a new technology. Historically, our quest for "progress" in agriculture and other human endeavors has had some inadvertent and adverse effects on the Earth's environment. Plowing ahead without adequate review of specific biotechnological projects and anticipating how they might affect our common future is not in our best interest.

Some environmentalists and more and more of the lay public, on the other hand, have assumed the role of a fearful, conservative mother who would just as soon keep the baby in its crib, unrealistically trying to prevent it from growing up at all. What's more, they're smelling smoke in reports of super weeds and contaminated taco shells and wondering whether the baby is playing with matches. And, at what therefore seems to be a critical time, the dysfunctional parents are having great difficulty communicating with each other.

A new agricultural innovation, biotechnology, has been born. This baby's potential is undeniable. Vitamin A–providing "golden" rice, developed without commercial support so that the seed might be given away to Third World farmers,[48] is a good example of that potential. But it is time, as a society, to utilize the facts at hand and our historical perspective to select an effective, transparent set of parenting guidelines and realistically face the challenge of rearing this brilliant child.

End Notes

Preface

1. From The Evolution of Physics, as quoted in Mather, R., *A Garden of Unearthly Delights: Bioengineering and the Future of Food.* New York: Dutton, 1995, pp. 17–18.

Chapter One

1. Owen, M. March 10, 1988. "Davis firms branching out in biotechnology." *Davis Enterprise.*
2. Cony, A. August 16, 1988. "Calgene short-circuits tomato rot." *Sacramento Bee.*
3. Goll, D. August 16, 1988. "Calgene claims breakthrough that will lead to 'supertomato.'" *Sacramento Union.*
4. Cony, op. cit.
5. Ibid.
6. Goll, op. cit.
7. Ibid.
8. Cony, op. cit.
9. Salquist, R. Letter to shareholders. In: Calgene, Inc., 1988 Annual Report.
10. Cony, op. cit.
11. Gore, A. 1992. *Earth in the Balance: Ecology and the Human Spirit.* New York: Penguin Books, p. 140.
12. Calgene, Inc., 1987 Annual Report, p. 16.
13. Goll, op. cit.
14. *Webster's New Collegiate Dictionary.* 1981. Springfield, MA: Merriam-Webster.
15. Goll, op. cit.
16. Ibid.
17. Salquist, op. cit.
18. Galbraith, J. K. 1996. *The Good Society: The Humane Agenda,* p. 37. New York: Houghton Mifflin.
19. Goll, op. cit.

End Notes

Chapter Two

1. Rabinow, P. 1996. *Making PCR: A Story of Biotechnology*, p. 25. Chicago: University of Chicago Press.
2. Ibid., p. 12.
3. Hobson, G. 1965. "The firmness of tomato fruit in relation to polygalacturonase activity." *Journal of Horticultural Science*, 40: 66–72.
4. Slater, A., M. J. Maunders, K. Edwards, W. Schuch, and D. Grierson. 1985. "Isolation and characterization of cDNA clones for tomato polygalacturonase and other ripening related proteins." *Plant Molecular Biology*, 5: 137–147.
5. Sheehy, R. E., J. Pearson, C. J. Brady, and W. R. Hiatt. 1987. "Molecular characterization of tomato fruit polygalacturonase." *Molecular and General Genetics*, 208: 30–36.
6. DellaPenna, D., D. C. Alexander, and A. B. Bennett. 1986. "Molecular cloning of tomato fruit polygalacturonase: analysis of polygalacturonase mRNA levels during ripening." *Proceedings of the National Academy of Sciences of the United States of America*, 83: 6420–6424.
7. Sheehy et al., op. cit.
8. Rabinow, op. cit., p. 26.
9. DellaPenna et al., op. cit.
10. Sheehy et al., op. cit.
11. Fillatti, J. J., J. Kiser, R. Rose, and L. Comai. 1987. "Efficient transfer of a glyphosate tolerance gene into tomato using a binary *Agrobacterium tumefaciens* vector." *Biotechnology*, 5: 726–730.
12. Nash, M. J. 1997. "The age of cloning." *Time*, March 10, pp. 62–65.
13. Ibid.
14. Smith, C., C. Watson, J. Ray, C. Bird, P. Morris, W. Schuch, and D. Grierson. 1988. "Antisense RNA inhibition of polygalacturonase gene expression in transgenic tomatoes." *Nature*, 334: 724–726.
15. Sheehy, R. E., M. Kramer, and W. R. Hiatt. 1988. "Reduction of polygalacturonase activity in tomato fruit by antisense RNA." *Proceedings of the National Academy of Sciences of the United States of America*, 85: 8805–8809.
16. Ibid.
17. Ibid.
18. Smith et al., op. cit.
19. Ibid.
20. Ibid.
21. Grierson, D. September 9, 1993. "What can we learn about physiology from plants transformed with antisense RNA?" Antisense Symposium, Iowa State University, Ames, IA.
22. Giovannoni, J. J., D. DellaPenna, A. B. Bennett, and R. L. Fischer. 1989. "Expression of a chimeric polygalacturonase gene in transgenic rin (ripening inhibitor) tomato fruit results in polyuronide degradation but not fruit softening." *Plant Cell*, 1: 53–63.
23. Smith et al., op. cit.
24. Schuch, W., G. Hobson, J. Kanczler, G. Tucker, N. Robertson, D. Gri-

erson, S. Bright, and C. Bird. 1991. "Fruit quality characteristics of transgenic tomato fruit with altered polygalacturonase activity." *Horticultural Science*, 26: 1517–1520.

25. Kramer, M., R. A. Sanders, R. E. Sheehy, M. Melis, M. Kuehn, and W. R. Hiatt. 1990. "Field evaluation of tomatoes with reduced polygalacturonase by antisense RNA." In: *Horticultural Biotechnology*, A. Bennett and S. O'Neill, eds. New York: Wiley-Liss, pp. 347–355.
26. Ibid.
27. Ibid.
28. Ibid.
29. Ibid.
30. Slater et al., op. cit.
31. DellaPenna et al., op. cit.
32. Grierson, D., G. A. Tucker, J. Keen, J. Ray, C. R. Bird, and W. Schuch. 1986. "Sequencing and identification of a cDNA clone for tomato polygalacturonase." *Nucleic Acids Research*, 14: 8595–8603.
33. Hiatt, W. R., R. E. Sheehy, C. K. Shewmaker, J. C. Kridl, and V. Knauf. 1989. "PG gene and its use in plants," U.S. Patent no. 4,801,540.
34. Kramer, M., R. Sanders, H. Bolkan, C. Waters, R. E. Sheehy, and W. R. Hiatt. 1992. "Postharvest evaluation of transgenic tomatoes with reduced levels of polygalacturonase: processing, firmness and disease resistance." *Postharvest Biology and Technology*, 1: 241–255.
35. Grierson, op. cit.
36. Grierson, D., and W. Schuch. 1993. "Control of ripening." *Philosophical Transactions of the Royal Society of London*, 342: 241–250.
37. Gray, J. E., S. Picton, J. J. Giovannoni, and D. Grierson. 1994. "The use of transgenic and naturally occurring mutants to understand and manipulate tomato fruit ripening." *Plant, Cell and Environment*, 17: 557–571.
38. Martin, R. M. October 17, 1992. "Host susceptibility genes: tomato polygalacturonase." CEPRAP Second Annual Retreat, Fallen Leaf Lake, CA.
39. Shewmaker, C. K., J. C. Kridl, W. R. Hiatt, and V. Knauf. 1992. "Antisense regulation of gene expression in plant cells," U.S. Patent no. 5,107,065.

Chapter Three

1. Pollan, M. October 25, 1998. "Playing God in the garden." *New York Times Magazine*.
2. Salquist, R. Letter to shareholders. Calgene, Inc., 1990 Annual Report.
3. Raphals, P. 1990. "Does medical mystery threaten biotech?" *Science*, 249: 619; Mayeno, A. N., and G. J. Gleich. 1994. "Eosinophilia-myalgia syndrome and tryptophan production: a cautionary tale." *Trends in Biotechnology*, 12: 346–352.
4. Salquist, op. cit.
5. *Federal Register*, May 1, 1991, 58(84): 20004.

End Notes

6. Beck, E. 1982. "Nucleotide sequence and exact localization of the neomycin phosphotransferase gene from transposon Tn5." *Gene*, 19: 327–336.
7. Ibid.
8. Davies, J., and D. I. Smith. 1978. "Plasmid-determined resistance to antimicrobial agents." *Annual Review of Microbiology*, 32: 469–518; Davies, J. E. 1986. "Aminoglycoside-aminocyclitol antibiotics and their modifying enzymes." In: *Antibiotics in Laboratory Medicine*, 2nd ed., V. Lorian, ed. Baltimore: Williams & Wilkins, pp. 791–809.
9. Beck, op. cit.
10. Calgene, Inc. 1990. "Request for advisory opinion, *kan*^{r} gene: safety and use in the production of genetically engineered plants," U.S. Food and Drug Administration docket no. 90A-0416, vol. 1, Food Additive Petition, FAP 3A4346, p. 163.
11. Ibid., p. 233.
12. Ibid., p. 234.
13. Ibid.
14. Ibid., p. 136.
15. Ibid., p. 111 and elsewhere.
16. Lehninger, A. L. 1975. *Biochemistry*, 2nd ed. New York: Worth, pp. 322–323.
17. Calgene, Inc., op. cit., p. 111.
18. Ibid., pp. 112–113.
19. Rissler, J., and M. Mellon. 1996. *The Ecological Risks of Engineered Crops*. Cambridge, MA: MIT Press, p. 69.
20. Stewart, G. J. 1989. "The mechanism of natural transformation." In: *Gene Transfer in the Environment*, S. B. Levy and R. V. Miller, eds. New York: McGraw-Hill, pp. 139–164.
21. Guyton, A. C. 1986. *Textbook of Medical Physiology*, 7th ed. Philadelphia: W. B. Saunders.
22. Ibid.
23. Pao, E. M., K. H. Fleming, P. M. Guenther, and S. J. Mickle. 1982. "Foods commonly eaten by individuals: amount per day and per eating occasion," Consumer Nutrition Center, Human Nutrition Information Service, Home Economics Research Report no. 44. Washington, DC: U.S. Department of Agriculture.
24. Ibid.
25. Atkinson, B. A. 1986. "Species incidence and trends of susceptibility to antibiotics in the United States and other countries: MIC and MBC." In: *Antibiotics in Laboratory Medicine*, 2nd ed., V. Lorian, ed. Baltimore: Williams & Wilkins.
26. *United States Pharmacopeia Dispensing Information*, 9th ed. 1989. "Drug information for the health care professional," vol 1A. Harrisonburg, VA: George Banta, p. 213.
27. Levy, S. B. 1984. "Antibiotic-resistant bacteria in food of man and animals." In: *Antimicrobials in Agriculture*, M. Woodbine, ed. London: Butterworth, pp. 525–531.
28. Armstrong, J. L., D. S. Shigeno, J. J. Calomiris, and R. J. Seidler. 1981.

End Notes

"Antibiotic-resistant bacteria in drinking water." *Applied and Environmental Microbiology*, 42: 277–283; Calomiris, J. J. 1984. "Association of metal tolerance with multiple antibiotic resistance of bacteria isolated from drinking water." *Applied and Environmental Microbiology*, 47: 1238–1242.

29. Calgene, Inc., op. cit., p. 205.
30. Goldburg, R., W. Baur, and M. Bhattacharya. July 30, 1991. Letter to FDA regarding a Calgene, Inc., request for an advisory opinion on use of *kan*^{r} in the production of genetically engineered tomato, cotton, and rapeseed plants, docket no. 90A-0416, p. 4.
31. Ibid., p. 5.
32. Calgene, Inc., op. cit., p. 307.
33. Henschke, R. B., and F. R. J. Schmidt. 1990. "Plasmid mobilization from genetically engineered bacteria to members of the indigenous soil microflora in situ." *Current Microbiology*, 20: 105–110.
34. Calgene, Inc., op. cit., p. 308.
35. Ibid., p. 304.
36. Ibid., p. 299.
37. Ibid., p. 301.
38. Ibid., p. 303.
39. Rick, C. M. 1949. "Rates of natural cross-pollination of tomatoes in various localities in California as measured by the fruits and seed set on male-sterile plants." *Proceedings of the American Society of Horticultural Science*, 52: 237–252; Rick, C. M., M. Holle, and R. W. Thorp. 1978. "Rates of cross-pollination in *Lycopersicon pimpinellifolium*: impact of genetic variation in floral characters." *Plant Systemics and Evolution*, 129: 31–44.
40. Simpson, D. M. 1954. "Natural cross-pollination in cotton," USDA Technical Bulletin no. 1094. Washington, DC: U.S. Department of Agriculture.
41. Olsson, G. 1960. "Species crosses within the genus *Brassica:* II. Artificial *Brassica napus* L." *Hereditas*, 46: 351–386.
42. Calgene, Inc., op. cit., p. 303.
43. Ibid., p. 304.
44. Ibid., p. 304.
45. Ibid., p. 298.
46. Stotzky, G., and H. Babich. 1986. "Survival of, and genetic transfer by, genetically engineered bacteria in natural environments." *Advances in Applied Microbiology*, 31: 93–137.
47. Schneider, K. January 27, 1990. "FDA ruling sought for engineered crops." *New York Times*.
48. Schumacher, J. November 21, 1998. "Big games have history of finishing on wild note." *Sacramento Bee*.

Chapter Four

1. Calgene, Inc. 1990. "Request for advisory opinion, *kan*^{r} gene: safety and use in the production of genetically engineered plants," U.S. Food

End Notes

and Drug Administration docket no. 90A-0416, vol. 1, Food Additive Petition, FAP 3A4346, p. 165.
2. Calgene, Inc. 1991. "Request for advisory opinion FLAVR SAVR™ tomato: status as food," U.S. Food and Drug Administration docket no. 91A-0330, p. 14.
3. Maryanski, J. December 13, 1999. "FDA policy: 1994 to the present." FDA Public Meeting: Biotechnology in the Year 2000 and Beyond, Oakland, CA.
4. Calgene, Inc., 1991, op. cit., p. 229.
5. Ibid., pp. 272–298.
6. Balter, M. 1997. "Transgenic corn ban sparks a furor," *Science*, 275: 1063.
7. Keith Redenbaugh, personal communication, 1999.
8. Calgene, Inc., 1991, op. cit., p. 339.
9. Ibid.
10. Ibid., p. 340.
11. Ibid., p. 341.
12. Southern, E. M. 1975. "Detection of specific sequences among DNA fragments separated by gel electrophoresis." *Journal of Molecular Biology*, 98: 503–517.
13. Calgene, Inc., 1991, op. cit., p. 347.
14. Jorgenson, R., C. Snyder, and J. D. G. Jones. 1987. "T-DNA is organized predominantly in inverted repeat structures in plants transformed with *Agrobacterium tumefaciens* C58 derivatives." *Molecular and General Genetics*, 207: 471–477.
15. Calgene, Inc., 1991, op. cit., pp. 318–326.
16. Ibid., p. 337.
17. Ibid., pp. 329–333.
18. Rick, C. M. 1978. "The Tomato." *Scientific American*, 239: 77–87.
19. Jones, D. D., and J. H. Maryanski. 1990. "Safety considerations in the evaluation of transgenic plants for human food." In M. A. Levin and H. S. Strauss, *Risk Assessment in Genetic Engineering*. McGraw-Hill, New York, p. 70.
20. Calgene, Inc., 1991, op. cit., p. 337.
21. Zitnak, A., and G. R. Johnston. 1970. "Glycoalkaloid content of B5141-6 potatoes." *American Potato Journal*, 47: 256.
22. Morris, S., and T. Lee. 1984. "The toxicity and teratogenicity of Solanaceae glycoalkaloids, particularly those of the potato (*Solanum tuberosum*): a review." *Food Technology in Australia*, 36: 118–124.
23. Calgene, Inc., 1991, op. cit., p. 338.
24. Ibid., pp. 337–338.
25. Ibid., pp. 564–565.
26. Bird, C. R., C. J. S. Smith, J. A. Ray, P. Moureau, M. W. Bevan, A. S. Bird, S. Hughes, P. C. Morris, D. Grierson, and W. Schuch, 1988. "The tomato polygalacturonase gene and ripening-specific expression in transgenic plants." *Plant Molecular Biology*, 11: 651–662.
27. Calgene, Inc., 1991, op. cit., pp. 390–392.
28. Ibid.
29. Mellon, M. July 30, 1991. National Wildlife Federation comments to

the Food and Drug Administration on a Calgene, Inc., request for an advisory opinion on the use of a kanamycin resistance gene in the production of genetically engineered food crops, p. 5.

30. Calgene, Inc., press release, August 12, 1991. "Calgene requests FDA review of FLAVR SAVR™ tomato."

Chapter Five

1. Cony, A. September 27, 1993. "Calgene's tomato to leap from lab into salad bowl." *Sacramento Bee*.
2. Week, L., and J. M. Berry. April 6, 1997. "Up close and personal with the second most powerful American." *Washington Post (Sacramento Bee)*.
3. Strunk, W., Jr., and E. B. White. 1959. *The Elements of Style*. New York: Macmillan, p. 69.
4. Week and Berry, op. cit.
5. Ibid.
6. Parietti, J. 1998. *The Book of Truly Stupid Business Quotes*. New York: HarperBusiness.
7. Pendick, Daniel. 1992. "Better than the real thing." *Science News*, 142: 376–377.
8. Roussel, P., K. Saad, and T. Erickson. 1991. *Third Generation R&D*. Cambridge, MA: Harvard University Business School Press.
9. Hamilton, J. O., J. Carey, and J. F. Siler. March 2, 1992. "The country cousin is blossoming, too." *Business Week*.
10. Ibid.
11. Poletti, T. March 1, 1992. "Ag-biotech hopeful after Bush boost." *San Francisco Examiner*.
12. Hamilton et al., op. cit.
13. Adler, J., and L. Denworth. March 9, 1992. "Flavr Savr." *Newsweek*.
14. Redenbaugh, K., W. Hiatt, B. Martineau, M. Kramer, R. Sheehy, R. Sanders, C. Houck, and D. Emlay. 1992. *Safety Assessment of Genetically Engineered Fruits and Vegetables: A Case Study of the Flavr Savr™ Tomato*. Boca Raton, FL: CRC Press.
15. Shewmaker, C. K., J. C. Kridl, W. R. Hiatt, and V. Knauf. 1992. "Antisense regulation of gene expression in plant cells," U.S. Patent no. 5,107,065.
16. *Federal Register*, May 29, 1992, 57(104): 22984–23005.
17. Maryanski, J. H. October 7, 1999. Statement before the Senate Committee on Agriculture, Nutrition and Forestry, U.S. Senate.
18. *Federal Register*, op. cit., p. 22985.
19. Ibid., pp. 22989–22990.
20. Ibid., p. 22986.
21. Ibid., p. 23004.
22. Ibid., p. 22987.
23. Ibid., p. 23004.
24. Kessler, D. A., M. R. Taylor, J. H. Maryanski, E. L. Flamm, and L. S. Kahl. 1992. "The safety of foods developed by biotechnology." *Science*, 256: 1747–1832.

25. *Federal Register,* op. cit., p. 22985.
26. Crouse, G. B. 1992. "Regulators request for comments on Calgene's 'Flavr Savr' tomato draws little response, negative or positive." *Biotech Daily,* 1(15): 1–2.
27. Ibid., p. 2.
28. Cony, op. cit.
29. Zambryski, P. C. 1992. "Chronicles from the *Agrobacterium*–plant cell DNA transfer story." *Annual Review of Plant Physiology and Plant Molecular Biology,* 43: 465–490.
30. Cony, A. March 3, 1992. "Panel says Flavr Savr tomato safe." *Sacramento Bee.*
31. Ibid.
32. Cony, A. November 30, 1992. "Selling the public on a super tomato." *Sacramento Bee.*
33. Cole, J. August 8, 1992. "3 shippers take plunge into biotech." *The Packer.*
34. Ibid.
35. DNA Plant Technology, Inc., press release, December 10, 1992. "DNAP's VineSweet tomato to be grown by Meyer Tomatoes: vine-ripen variety has superior taste and longer shelf life."
36. Ibid.
37. Cony, September 27, 1993, op. cit.
38. Barnum, A. September 30, 1993. "Calgene has tomato industry seeing red." *San Francisco Chronicle.*
39. Barnum, A. May 19, 1994. "FDA gives green light to souped-up tomato." *San Francisco Chronicle.*

Chapter Six

1. Cony, A. April 6, 1994. "FDA scientists find Flavr Savr safe." *Sacramento Bee.*
2. Stoddard, M. N. November 1995. "Exclusive interview: consumer information on aspartame." *Nutrition & Healing.*
3. Cony, op. cit.
4. Jukes, T. H. 1988. "Hazards of biotechnology: facts and fancy." *Journal of Chemical Technology and Biotechnology,* 43: 1–11.
5. Walter, B. May 26, 1994. "Calgene has momentary monopoly." *Sacramento Bee.*
6. McClatchy News Service. May 14, 1994. "Flavr-Savr delay costly to Calgene." *Oakland Tribune.*
7. Calgene, Inc., press release, May 13, 1994. "Calgene announces third quarter results."
8. McClatchy News Service, op. cit.
9. Ibid.
10. Calgene, Inc., op. cit.
11. Ibid.
12. Barnum, A. May 19, 1994. "Biotech tomato wins final OK for marketing." *San Francisco Chronicle.*

13. Leary, W. May 19, 1994. "F.D.A. approves altered tomato that will remain fresh longer." *New York Times.*
14. Pollan, M. October 25, 1998. "Playing God in the garden." *New York Times Magazine.*
15. Barnum, op. cit.
16. Elmer-Dewitt, P. May 30, 1994. "Fried gene tomatoes." *Time.*
17. Leary, op. cit.
18. Barnum, op. cit.
19. Ibid.
20. Jenkins, N. December 1994. "Retail revolution or produce footnote?" *Produce Merchandising.*
21. Ibid.
22. Ibid., p. 26.
23. Martineau, B., T. A. Voelker, and R. A. Sanders. 1994. "On defining T-DNA," *Plant Cell*, 6: 1032–1033.
24. Barnum, op. cit.
25. Saekel, K. May 19, 1994. "What the tasters say." *San Francisco Chronicle.*
26. Cony, A. September 27, 1993. "Calgene's tomato to leap from lab into salad bowl." *Sacramento Bee.*
27. Sugarman, C. June 8, 1994. "Tasting . . . 1, 2, 3, tasting." *Washington Post.*
28. Barnum, op. cit.
29. Elmer-Dewitt, op. cit.
30. O'Neill, M. May 19, 1994. "No substitute for summer." *New York Times.*
31. O'Neill, M. May 19, 1994. "Flavr Savr serves, but it's no beefsteak." *Davis Enterprise (New York Times* News Service).
32. Walter, op. cit.
33. Ibid.
34. Ibid.
35. Ibid.
36. Ibid.
37. Calgene, Inc., 1994 Annual Report, p. 20.
38. Ibid., p. 21.
39. Ibid., p. 6.
40. Anonymous. November 9, 1994. "Calgene loses $9.5 million, *Sacramento Bee.*
41. Calgene, Inc., press release, November 8, 1994. "Calgene announces first quarter results."
42. Jenkins, op. cit.
43. Anonymous. November 18, 1994. "Another patent for Calgene." *Sacramento Bee.*
44. Anonymous. November 26, 1994. "At last a tomato with home-grown garden flavor." *Sacramento Bee.*
45. Jenkins, op. cit.
46. Anonymous, November 18, 1994, op. cit.
47. Jenkins, op. cit.

End Notes

Chapter 7

1. Unrein, J. March 24, 1995. "Firms preparing to unveil modified tomato," *The Packer.*
2. Calgene, Inc., press release, January 10, 1995. "Calgene announces expansion of MacGregor's™ tomato packing and distribution network and limited availability of tomatoes this winter."
3. Unrein, op. cit.
4. Ibid.
5. Ibid.
6. Greenberg, H. January 15, 1995. "Calgene's biotech bounty disappears from grocers' shelves." *San Francisco Chronicle.*
7. Ibid.
8. Ibid.
9. Calgene, Inc., 1994 Annual Report, p. 6.
10. Greenberg, op. cit.
11. Walter, B. March 10, 1995. "'Shorts' take it in shorts." *Sacramento Bee.*
12. Bullock, W. O., Jr. January 1995. "Ag-biotech forecast." In: *Agricultural Biotechnology Notes.* Office of Ag Biotech, USDA.
13. Beck, H. February 6, 1995. "Layoffs are part of industry risk, Calgene staff say." *Davis Enterprise.*
14. Calgene, Inc., press release, January 31, 1995.
15. Ibid.
16. Canine, C. January/February 1991. "A matter of taste: who killed the flavor in America's supermarket tomatoes?" *Eating Well.*
17. Seabrook, J. July 19, 1993. "Tremors in the hothouse." *The New Yorker.*
18. Ibid.
19. Parrish, M. October 16, 1994. "Innovative produce company takes financial battering." *Los Angeles Times.*
20. Ibid.
21. Ibid.
22. Unrein, op. cit.
23. Ibid.
24. Ibid.
25. Greenberg, op. cit.
26. Anonymous. May 17, 1995. "Calgene hit with setback." *Sacramento Bee.*
27. Walter, B. April 4, 1995. "Calgene, arch foe ready for biotech court battle." *Sacramento Bee.*
28. Bernstein, K. May 1, 1995. "Titillating story diverts focus of antisense trial." *BioCentury.*
29. Ibid.
30. Ibid.
31. Ibid.
32. King, R. T., Jr. April 24, 1995. "Expert calls Calgene research on gene-altering method flawed." *Wall Street Journal.*
33. Ibid.

34. Ibid.
35. Bernstein, op. cit.
36. King, op. cit.
37. Ibid.
38. Walter, op. cit.
39. Anonymous, May 17, 1995, op. cit.
40. Black, J. May 17, 1995. "Genetically altered tomatoes ripe for tossing in Seattle salads." *Seattle Times.*
41. Ibid.
42. MacPherson, K. May 17, 1995. "Calgene Flavr Savr beats the standard varieties in a blind taste test." *Star-Ledger*, Newark, N.J.
43. Ibid.
44. Ibid.
45. Schnitt, P. April 25, 1995. "Calgene stewing over testimony." *Sacramento Bee.*
46. Anonymous, May 17, 1995, op. cit.
47. Ibid.
48. Brammer, R. May 22, 1995. "Buying and selling biotech stocks: the right formula: an interview with Mark Lampert." *Barron's.*
49. Ibid.
50. Walter, B. May 24, 1995. "Calgene stock bounces back after tumble." *Sacramento Bee.*
51. Ibid.
52. Ibid.
53. Anonymous. June 2, 1995. "Merger rumor boosts Calgene." *Sacramento Bee.*
54. Ibid.
55. Craig, C. June 29, 1995. "Monsanto to buy 50 percent stake in Calgene." *BioWorld Today.*
56. Ibid.
57. Walter, B. June 29, 1995. "Monsanto buys piece of Calgene." *Sacramento Bee.*
58. Bernstein, K. June 29, 1995. "Calgene-Monsanto tie knot." *BioCentury Extra.*
59. Ibid.
60. Walter, June 29, 1995, op. cit.
61. Bernstein, op. cit.
62. Walter, June 29, 1995, op. cit.
63. Bernstein, op. cit.
64. Ibid.
65. Walter, June 29, 1995, op. cit.
66. Anonymous. June 1995. "Flavr Savr goes nationwide." *The California Tomato Grower.*
67. Bernstein, op. cit.
68. Walter, June 29, 1995, op. cit.
69. Ibid.
70. Anonymous. November 14, 1995. "Prices, supply hurt Calgene." *Sacramento Bee.*

71. Hubert, C. June 30, 1995. "Bomber mails out renewed demands." *Sacramento Bee.*
72. Steyer, R. September 18, 1995. "In search of the perfect bioengineered tomato." *St. Louis Post-Dispatch.*
73. Ibid.
74. Canine, op. cit.
75. Ibid., p. 46.
76. Anonymous. March 2, 1996. "Calgene targets UK for tomatoes." *Sacramento Bee.*
77. Calgene, Inc., press release, October 25, 1994. "Calgene begins world's first commercial planting of genetically engineered plant oil."
78. Walter, B. February 3, 1996. "Calgene wins lawsuit over biotech research." *Sacramento Bee.*
79. Anonymous, March 2, 1996, op. cit.
80. Calgene, Inc., press release, June 23, 1995. "Calgene requests clearance for Flavr Savr tomato in United Kingdom."
81. Calgene, Inc., press release, March 1, 1996. "Calgene receives UK food safety approval for Flavr Savr™ tomato."
82. Anonymous, November 14, 1995, op. cit.
83. Anonymous. February 7, 1996. "Calgene reports loss." *Sacramento Bee.*
84. Glover, M. April 30, 1997. "Calgene is going out with a bang." *Sacramento Bee.*
85. Ibid.
86. Groves, M. August 1, 1996. "Monsanto to raise stake in Calgene and replace CEO." *Los Angeles Times.*
87. Gomes, L. August 1, 1996. "Calgene CEO quits as Monsanto Co. acquires control." *Wall Street Journal.*
88. Calgene, Inc., press release, July 31, 1996. "Calgene announces planned $50 million equity investment by Monsanto."
89. Groves, op. cit.
90. Hall, C. T. August 1, 1996. "Calgene chief exec bows out." *San Francisco Chronicle.*
91. Glover, M. November 14, 1996. "Monsanto gains 55% of Calgene." *Sacramento Bee.*
92. Anonymous. January 29, 1997. "Monsanto offers $30 million for rest of Calgene." *New York Times.*
93. Glover, M. January 29, 1997. "Monsanto bids for Calgene." *Sacramento Bee.*
94. Glover, M. April 30, 1997, op. cit.
95. Lane, R. April 1, 1997. "Monsanto raises bid for remaining Calgene stock." *Davis Enterprise.*
96. Glover, M. April 2, 1997. "Monsanto finishes Calgene takeover." *Sacramento Bee.*
97. Glover, M. September 24, 1996. "Calgene and Tokyo firm near genetic research pact." *Sacramento Bee.*
98. Ibid.

End Notes

99. Young, E. July 30, 1994. "Altered tomato faces ban from British shops." *New Scientist*, 143:9.
100. Swett, C. July 24, 1995. "Calgene non-rival gets OK for altered tomato." *Sacramento Bee.*
101. Stecklow, S. October 26, 1999. "'Genetically Modified' on the label means . . . well, it's hard to say." *Wall Street Journal.*

Epilogue

1. Doyle, M. February 23, 2000. "Boxer: label altered crops." *Sacramento Bee.*
2. Miller, H. February 28, 2000. "Food labels would be antibiotech." *Sacramento Bee.*
3. Cardello, C. June 21, 2000. "Calgene sabotage attempt probed." *Davis Enterprise.*
4. Ibid.
5. Barboza, D. November 18, 1999. "2 sides square off on genetically altered food." *New York Times.*
6. Kasler, D. May 10, 2000. "Biotech foods get Safeway yes vote." *Sacramento Bee.*
7. Anonymous. March 27, 2000. "Boston biotech protest stays peaceful." *Boston Globe (Sacramento Bee).*
8. Weiss, R. May 3, 2000. "Plan to set rules for genetically altered food." *Washington Post (Sacramento Bee).*
9. Archer Daniels Midland press release, August 31, 1999. "ADM statement to suppliers regarding genetically enhanced crops."
10. Weiss, R. September 26, 1999. "Firms make farmers pay for rising distrust of gene-altered food." *Washington Post (Sacramento Bee).*
11. Charter, D., and V. Elliott. June 7, 2000. "Princess fuels split on GM food." *London Times.*
12. Kasler, D. June 25, 2000. "Biotech backlash." *Sacramento Bee.*
13. Miller, H. 1994. "Risk assessment experiments and the new biotechology." *Trends in Biotechnology,* 12: 292–295.
14. Kasler, op. cit.
15. Brand, W. December 14, 1999. "Genetically altered produce sparks protest on labeling." *Oakland Tribune.*
16. Lovett, R. June 20, 1999. "Warning from the butterflies." *Sacramento Bee.*
17. Ibid.
18. Hotz, R. March 8, 1996. "Pesticide-proof plants let weeds in on their secret." *Los Angeles Times (Sacramento Bee).*
19. Specter, M. April 10, 2000. "The pharmageddon riddle." *The New Yorker.*
20. Fox, J. 1990. "Everything but the animals." *Biotechnology,* 8:822.
21. Mann, C. 1999. "Biotech goes wild." *Technology Review,* 102:36-40.
22. Doyle, M. April 6, 2000. "Lofty panel backs gene-altered foods." *Sacramento Bee.*

23. Pollan, M. October 25, 1998. "Playing God in the garden." *New York Times Magazine.*
24. Ibid.
25. Mann, op. cit.
26. Brasher, P. April 2, 2000. "Glickman asks panel to review 'terminator' seeds." Associated Press (*Davis Enterprise*).
27. Weiss, op. cit.
28. Shapiro, L., M. Hager, P. Wingert, and K. Springen. June 6, 1994. "A tomato with a body that just won't quit." *Newsweek.*
29. Fox, op. cit.
30. Ibid.
31. McHughen, A. 2000. *Pandora's Picnic Basket.* Oxford: Oxford University Press, p. 151.
32. Doyle, M. May 4, 2000. "Biotech food rules too lax, critics say." *Sacramento Bee.*
33. Jansen, B. February 23, 2000. "Boxer's labeling bill faces foes." Associated Press (*Davis Enterprise*).
34. McHughen, op. cit., pp. 201–229.
35. Barboza, D. June 4, 2000. "Modified foods put companies in a quandary." *New York Times.*
36. Kasler, D. October 13, 2000. "Safeway pulls 2 taco shell brands." *Sacramento Bee.*
37. Doyle, op. cit.
38. McHughen, op. cit., p. 151.
39. Ibid., pp. 201–229.
40. Weiss, op. cit.
41. Barboza, op. cit.
42. Koenig, D. February 1, 2000. "Frito-Lay asks farmers not to use lab-bred corn." Associated Press (*Davis Enterprise*).
43. Anonymous. April 29, 2000. "No change, Kellogg says." *Sacramento Bee.*
44. Doyle, M. February 1, 2001. "Clearer biotech rules sought." *Sacramento Bee.*
45. Brasher, P. June 30, 2000. "Farmers planting fewer crops of gene-altered corn." Associated Press (*Davis Enterprise*).
46. Doyle, op. cit.
47. Hopkins, D., R. Goldburg, and S. Hirsch, 1991. *A Mutable Feast: Assuring Food Safety in the Era of Genetic Engineering.* New York: Environmental Defense Fund.
48. Hotz, R. January 14, 2000. "Scientists alter rice for vitamin A boost." *Los Angeles Times (Sacramento Bee).*

Index

Index

Index

About The Author

Belinda Martineau earned her A.B. in biology from Harvard College and her Ph.D. in genetics from U.C. Berkeley. Prior to joining the staff at Calgene, Inc. in 1988 she was a post-doctoral fellow in the Department of Molecular Genetics and Cell Biology at the University of Chicago. She has published over two dozen articles in scientific journals and has been issued patents on four inventions that utilize genetic engineering. Despite her molecular biological background, Dr. Martineau's perspective on biotech food has been called "emphatically nonpartisan." She lives in Davis, California, with her husband and two children.